Construction of Highly Ordered Nanomaterials Composed of Protein Containers and Plasmonic Nanoparticles

Construction of Highly Ordered Nanomaterials Composed of Protein Containers and Plasmonic Nanoparticles

Dissertation

zur Erlangung der Würde des Doktors der Naturwissenschaften der Fakultät für Mathematik, Informatik und Naturwissenschaften, Fachbereich Chemie der Universität Hamburg

vorgelegt von

Marcel Josef Lach

aus Düsseldorf

Hamburg 2020

Bibliografische Information der Deutschen Nationalbibliothek
Die Deutsche Nationalbibliothek verzeichnet diese Publikation in der
Deutschen Nationalbibliografie; detaillierte bibliographische Daten sind im Internet
über http://dnb.d-nb.de abrufbar.
1. Aufl. - Göttingen: Cuvillier, 2020
Zugl.: Hamburg, Univ., Diss., 2020

Gutachter der Dissertation	Prof. Dr. Tobias Beck
	Prof. Dr. Alf Mews
Gutachter der Disputation	Prof. Dr. Tobias Beck
	Prof. Dr. Christian Betzel
	Prof. Dr. Axel Jacobi von Wangelin
	Prof. Dr. Holger Lange
Tag der Disputation	25.09.2020
Druckfreigabe	20.10.2020

Nonnenstieg 8, 37075 Göttingen
Telefon: 0551-54724-0
Telefax: 0551-54724-21
www.cuvillier.de

1. Auflage, 2020
Gedruckt auf umweltfreundlichem, säurefreiem Papier aus nachhaltiger Forstwirtschaft.

ISBN 978-3-7369-7296-4
eISBN 978-3-7369-6296-5

To my family

Abstract

Nanoparticles are interesting building blocks for the construction of new nanomaterials. By controlling the composition as well as the structure of these nanoparticle-based materials, novel properties can emerge. Currently, major challenges in nanoparticle material fabrication are low-order assembly, small domain sizes and long interparticle distances. These limitations will be overcome by using protein containers as an atomically precise ligand shell.

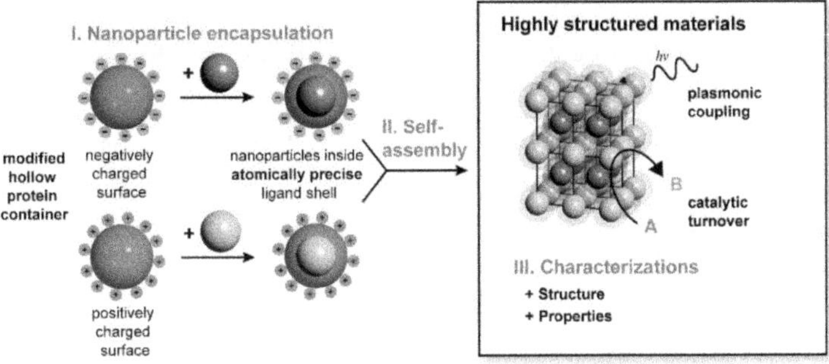

A new type of protein-based material was realized by using an innovative design approach with two oppositely charged protein containers as building blocks. Nanoparticle incorporation inside the protein container cavity was performed with two different methods. On the one hand, gold nanoparticles were synthesized and encapsulated into the cavity of the protein container by a dis- and reassembly approach. On the other hand, metal oxide nanoparticles were synthesized *in situ* inside the protein container cavity. Highly ordered nanoparticle superlattices were generated by self-assembly of nanoparticle-loaded oppositely charged protein container *via* electrostatic interactions into three-dimensional binary crystals. The hybrid materials possess a high nanoparticle content and display short interparticle distances, which is essential for emergent properties based on coupling between the particles. Structural studies and characterization of optical as well as catalytic properties of the binary crystals were performed. Because the protein scaffold is independent of the nanoparticle cargo, this modular approach will enable tuning of the material properties by choice of nanoparticle content, assembly type and protein container type.

Zusammenfassung

Nanopartikel sind ein interessanter Baustein für die Konstruktion von neuartigen Nanomaterialien. Durch gezieltes Einstellen der Zusammensetzung sowie der Struktur dieser nanopartikelbasierenden Materialien können neue Eigenschaften entstehen. Momentan sind die größten Herausforderungen in der Materialherstellung mit Nanopartikeln als Baustein die niedrige Ordnung, kleine Domaingrößen und der große interpartikulare Abstand. Diese Limitierungen werden durch den Einsatz von Proteincontainern als atomar präzise Ligandenhülle überwunden.

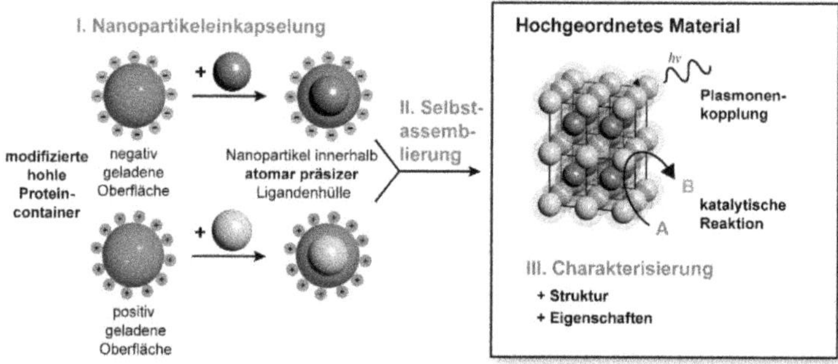

Eine neue Art von proteinbasierten Materialien wurde durch die Anwendung von einem innovativem Designansatz mit zwei entgegengesetzt geladenen Proteincontainern als Baustein verwirklicht. Die Nanopartikeleinlagerung innerhalb der Proteincontainerkavität wurde mit zwei Methoden durchgeführt. Zum einen wurden Goldnanopartikel synthetisiert und durch Zerlegen und Wiederaufbauen des Proteincontainers eingekapselt. Zum anderen wurden Metalloxidnanopartikel *in situ* in der Proteincontainerkavität synthetisiert. Hochgeordnete Nanopartikelsupergitter wurden durch Selbstanordnung der entgegengesetzt geladenen nanopartikelbeladenen Proteincontainern durch elektrostatische Wechselwirkungen zu einem dreidimensionalen binären Kristall aufgebaut. Das Hybridmaterial besitzt einen hohen Nanopartikelanteil und weist kurze interpartikulare Abstände auf, welche essenziell für die Entstehung von Eigenschaften basierend auf interpartikularen Wechselwirkungen sind. Es wurden strukturelle Untersuchungen und die Charakterisierung sowohl der optischen als auch der katalytischen Eigenschaften der binären Kristalle durchgeführt. Da das Proteingerüstes unabhängig von der Nanopartikelbeladung ist, wird dieser modulare Ansatz das Einstellen von Materialeigenschaften durch die Wahl des Nanopartikels, der Struktur des Proteingerüstes und des Proteincontainertyps ermöglichen.

The work presented in this thesis was carried out at the Institute of Inorganic Chemistry of the RWTH Aachen University between October 2016 and December 2019 and at the Institute of Physical Chemistry of the University Hamburg between January 2020 and September 2020. Funding in form of a PhD scholarship was received from the Cusanuswerk between October 2017 and September 2020.

Publications

Parts of this thesis have been published or are in the process of being published:

I. Lach, M.; Künzle, M.; Beck, T. Free-Standing Metal Oxide Nanoparticle Superlattices Constructed with Engineered Protein Containers Show in Crystallo Catalytic Activity. *Chem. Eur. J.* **2017**, *23*, 17482-17486.

II. Construction of Highly Ordered Materials Composed of Protein Containers and Plasmonic Nanoparticles, *in preparation*.

The author has also contributed to the following related publications:

III. Künzle, M.; Mangler, J.; Lach, M.; Beck, T. Peptide-Directed Encapsulation of Inorganic Nanoparticles into Protein Containers. *Nanoscale* **2018**, *10*, 22917-22926.

IV. Künzle, M.; Lach, M.; Beck, T. Crystalline Protein Scaffolds as a Defined Environment for the Synthesis of Bioinorganic Materials. *Dalton Trans.* **2018**, *47*, 10382-10387.

V. Lach, M.; Künzle, M.; Beck, T. Proteins as Sustainable Building Blocks for the Next Generation of Bioinorganic Nanomaterials. *Biochemistry* **2019**, *58*, 140-141.

VI. Künzle, M.; Lach, M.; Beck, T. Multi-Component Self-Assembly of Proteins and Inorganic Particles: From Discrete Structures to Biomimetic Materials. *Isr. J. Chem.* **2019**, *59*, 906-912.

VII. Künzle, M.; Lach, M.; Budiarta, M.; Beck, T. Highly-Order Structures Based on Molecular Interactions for the Formation of Natural and Artificial Biomaterials. *ChemBioChem* **2019**, *20*, 1637-1641.

VIII. Surface Charged Protein Containers as Tunable Matrix for the Assembly of Inorganic Nanoparticles into Crystalline Hybrid Materials, *in preparation*.

Conference contributions:

- Poster: "Controlling the Structure of Binary Protein Crystals Used for the Assembly of Inorganic Nanoparticle Superlattices." *NaNaX – Nanoscience with Nanocrystals*, Braga, Portugal **2017**.
- Talk: "Construction of Highly Ordered Materials Composed of Protein Containers and Nanoparticles." *Nanotage*, Karlsruhe, Germany **2017**.
- Poster: "Binary Protein Crystals for the Assembly of Inorganic Nanoparticle Superlattices." *26th Annual Meeting of the German Crystallographic Society*, Essen, Germany **2018**. *poster prize.*
- Talk and poster: "Binary Protein Crystals for the Construction of Highly Ordered Nanoparticle Superlattices." *3rd Meeting of the Young Crystallographers*, Aachen, Germany **2018**. *presentation and poster prize.*
- Poster: "Metal Oxide Nanoparticle Superlattices Constructed with Engineered Protein Containers." *19. Vortragstagung für Anorganische Chemie der Fachgruppen Wöhlervereinigung und Festkörperchemie und Materialforschung*, Regensburg, Germany **2018**. *poster prize.*
- Talk: "Construction of Highly Ordered Materials Composed of Protein Containers and Nanoparticles." *Nanotage*, Aachen, Germany **2019**.
- Poster: "Synthesis of Biohybrid Materials by the Assembly of Nanoparticle Superlattices with Engineered Protein Containers." *GRS/GRC Physical Virology*, Ventura, USA **2019**.
- Talk: "Construction of Highly Ordered Materials Composed of Protein Containers and Plasmonic Nanoparticles." *Joint Polish-German Crystallographic Meeting*, Wrocław, Poland **2020**.

Table of contents

1 **Introduction** **1**

2 **Theoretical background** **3**

2.1 Plasmonic nanoparticles 3

2.1.1 Synthesis and functionalization 3

2.1.2 Plasmonic nanoparticle properties and applications 7

2.2 Formation of nanoparticle superlattices 9

2.2.1 Self-assembly with organic ligands 9

2.2.2 Assembly with DNA molecules 12

2.3 Protein containers as a new building block 17

2.3.1 The ferritin protein container 19

2.3.2 Organization of ferritin into highly ordered structures 22

3 **Basis for the present work** **27**

3.1 Supercharging of the protein container ferritin 27

3.2 Nanoparticle synthesis 28

3.3 Self-assembly 29

4 **Concept of the work** **31**

5 **Results** **33**

5.1 Encapsulation of plasmonic nanoparticles 33

5.1.1 Gold nanoparticle cargo 34

5.1.1.1 Gold nanoparticle synthesis 35

5.1.1.2 Functionalization and characterization 38

5.1.1.3 Stability of the gold nanoparticles 48

5.1.2 Silver nanoparticle cargo 51

5.1.2.1 Silver nanoparticle synthesis 51

5.1.2.2 Functionalization and stability test 53

5.1.3 Dis- and reassembly of the protein container 56

5.1.3.1 Dis- and reassembly of $Ftn^{(pos)}$ 57

5.1.3.2 Dis- and reassembly of $Ftn^{(neg)}$ 60

5.1.4 AuNP encapsulation 65

5.1.4.1 Influence of the ligand shell 65

5.1.4.2 Influence of the ionic strength 69

5.1.4.3 Purification and optimization ... 71
5.1.4.3.1 Purification of $AuFtn^{(pos)}$ ... 72
5.1.4.3.2 Purification of $AuFtn^{(neg)}$ ... 75
5.1.5 AgNP encapsulation ... 79
5.1.5.1 Encapsulation in $Ftn^{(pos)}$ ... 79
5.1.5.2 Encapsulation in $Ftn^{(neg)}$ ... 80
5.2 Encapsulation of dye molecules ... 82
5.3 Synthesis of metal oxide nanoparticles *in situ* ... 85
5.4 Assembly of the building blocks ... 87
5.4.1 Batch crystallization ... 88
5.4.2 Formation of nanoparticle superlattices ... 91
5.5 Potential applications of the nanomaterial ... 97
5.5.1 Optical properties ... 97
5.5.2 Catalytic activity ... 101
6 Summary and perspective ... 109
7 Experimental Part ... 113
7.1 General ... 113
7.2 Chemicals ... 113
7.3 E-coli strains ... 113
7.4 Used buffers ... 114
7.5 Analytics ... 114
7.5.1 Transmission electron microscopy ... 114
7.5.2 Scanning electron microscopy ... 115
7.5.3 Dynamic light scattering ... 115
7.5.4 Zeta potential ... 115
7.5.5 Surface enhanced Raman spectroscopy ... 116
7.5.6 Atomic absorption spectroscopy ... 116
7.5.7 Nuclear magnetic resonance spectroscopy ... 116
7.5.8 UV-Vis spectroscopy ... 116
7.5.9 Optical microscopy ... 117
7.5.10 X-ray diffraction ... 117
7.5.11 Small angle X-ray scattering ... 117
7.5.12 Thermogravimetric analysis ... 118
7.5.13 Circular dichroism spectroscopy ... 120

7.5.14 Density functional theory calculations ... 120
7.5.15 Magnetic characterization ... 120
7.6 Production and purification of the ferritin variants ... 121
7.6.1 Positively charged protein container ... 121
7.6.2 Negatively charged protein container ... 122
7.7 Nanoparticle synthesis in the protein container ... 122
7.7.1 Synthesis of CeO_2 nanoparticles ... 122
7.7.2 Synthesis of FeO_x nanoparticles ... 123
7.8 Gold nanoparticle synthesis and functionalization ... 124
7.8.1 Leff synthesis ... 124
7.8.2 Peng synthesis ... 124
7.8.3 Calculation of gold nanoparticle concentration ... 125
7.8.4 Ligand exchange ... 125
7.9 Silver nanoparticle synthesis ... 126
7.9.1 Yamamoto synthesis ... 126
7.9.2 Hyeon synthesis ... 126
7.9.3 Ligand exchange ... 127
7.10 Nanoparticle stability tests ... 127
7.11 Dis- and reassembly ... 127
7.12 Encapsulation of plasmonic nanoparticles ... 128
7.13 Encapsulation of dyes ... 128
7.13.1 Rhodamine 6G in $Ftn^{(pos)}$ ... 128
7.13.2 Rhodamine B in $Ftn^{(neg)}$ ... 129
7.13.3 Determination of the dye loading ... 129
7.14 Catalysis with metal oxide nanoparticles in solution ... 129
7.14.1 Oxidase-like and peroxidase-like activity ... 129
7.14.2 Kinetic analysis ... 130
7.15 Kinetic measurements of the AuNPs in solution ... 130
7.16 Protein Crystallography ... 131
7.16.1 Hanging drop vapor diffusion ... 131
7.16.2 Batch crystallization ... 131
7.16.3 Stabilization of crystals ... 131
7.16.4 Catalytic activity ... 132

8 Appendix 133

9 Bibliography 153

10 List of abbreviations 171

11 List of chemicals 175

12 List of Figures 181

13 List of Tables 185

Acknowledgment - Danksagung 187

Curriculum Vitae 189

1 Introduction

Nanoparticles have become the focus of much attention due to their remarkable physical and chemical properties, which arise from their small size and high specific surface area.[1] They can be employed, among other applications, in catalysis,[2] imaging,[3,4] data storage[5] and drug-delivery[6] or for the production of new materials, such as magnetic fluids[7] and polymer nanocomposites.[8,9] The precise arrangement of nanoparticles in defined structures is a prerequisite for the generation of so-called metamaterials. These materials are envisioned to have emergent electronic, magnetic or optical properties because of the interactions between individual nanoparticles inside the materials.[10]

The formation of highly ordered materials requires discrete building blocks that carry the useful function and interact with each other in a structured manner.[11] Nanoparticles have remarkable properties, but are difficult to order due to their limited monodisperse size distribution. Control over the particle packing is strongly limited and only small domain sizes are obtained in binary structures.[12,13] As a consequence, the development of new methods for nanoparticle assembly are mandatory. Nature has always acted as role-model for the creation of new technologies. For example, the structure of bones has inspired mankind to construct buildings such as the Eiffel tower. In chemistry, the nature's principles to assemble fascinating complexes led to the establishment of supramolecular chemistry, where non-covalent interactions between molecules are utilized to create new materials.[14] With these design principles, the design and synthesis of molecular machines was possible.[15,16]

Several techniques have been developed to exploit biomolecules for the controlled nanoparticle assembly.[17,18] Typically, most work, pioneered by the MIRKIN and GANG groups, has focused on the assembly of oligonucleotide-functionalized nanoparticles into crystalline superlattices.[19,20] For instance, complementary DNA molecules were used as linkers between gold nanoparticles to create single-component and binary-component crystals.[21] In this way, the organization of various nanoparticle types is enabled to produce binary and ternary assemblies with tailorable crystal lattices and lattice parameters.[22,23] However, still several disadvantages exist such as the instability outside the growth medium or the low long-range order due to the polydispersity of the nanoparticles.

From a materials perspective, proteins are an interesting building block due to their atomically precise defined structure. Moreover, proteins can assemble via non-covalent bonds into multimeric complexes such as protein containers.[24] Protein containers are cage-like structures with an empty inner cavity, which can be used as nanoreactor for the synthesis of nanoparticles.[25] For example, the protein ferritin[26] has been extensively used for the size-constrained synthesis of nanoparticles including metals,[27-31] metal oxides[32–34] and semiconductor materials.[35–38] In previous work within this research group, these nanoparticle-protein container composites were assembled into a highly ordered structure by protein crystallization with the protein scaffold as matrix for the nanoparticle assembly.[39,40]

In this work, a novel strategy for the construction of biohybrid materials composed of plasmonic nanomaterials and protein containers is presented. Two oppositely charged ferritin protein containers are loaded with plasmonic nanoparticles using a new encapsulation approach by opening and closing the protein container while the nanoparticles are added. The crystallization of the building blocks leads to large crystalline assemblies with sizes up to hundred micrometers. Moreover, binary nanoparticle superlattices are generated because oppositely charged building blocks are utilized. These two building blocks enable the incorporation of one type of nanoparticle in each building block. Due to stabilization of the protein matrix with cross-linking agents, applications in catalysis or as optical materials are feasible.

2 Theoretical background

2.1 Plasmonic nanoparticles

The term nanoparticle is used for the associations of atoms with a dimension in the range of 1 nm to 100 nm, which classifies them between molecules and bulk materials.[41] The smaller size compared to bulk materials leads to novel properties that differ from the corresponding bulk material. These properties can be attributed to three different effects. First, the small nanoparticle size enhances surface-dependent properties because of the higher surface to volume ratio. A typical example is the higher catalytic activity due to increased number of catalytic active centers in comparison to the bulk material.[42] Second, size-dependent particle properties can emerge such as interference effects in photonic crystal[43] or superparamagnetic properties of Fe_3O_4 nanoparticles.[44] Third, size-dependent quantum effects occur caused by the influence of the nanoparticle size on the electronic structure. The transition from an atom with defined energy levels to the dispersed bands of collective ensembles of atoms leads to novel properties. For example, the position of the localize surface plasmon resonance (LSPR) of gold nanoparticles is redshifted with increasing nanoparticle size[45] or the luminescence of quantum dots is altered.[46] Plasmonic nanoparticles are metallic nanoparticles usually composed of noble metal atoms such as silver or gold. These particles are of high interest for the generation of novel nanomaterials because of their interesting size-dependent quantum effects. Structures based on plasmonic nanoparticles have potential for application in the construction of metamaterials with negative refractive index[47] and light-harvesting materials, such as in photovoltaics,[48] or sensing of biological analytes.[49]

2.1.1 Synthesis and functionalization

The application of nanoparticles in materials is strongly connected to the synthesis of nanoparticles in the favored size range. The requirements for the synthesis are a monodisperse size distribution, suppression of agglomeration processes and a specific control over the surface functionality. In general, nanoparticles can be obtained by two fundamental strategies: "top-down" or "bottom-up" approach. In the "top-down" approach, bulk material is minced to smaller particles for example in mechanical milling[50] or laser-ablation technique.[51] Industrially, "top-down" methods are favored due to the economical production of large amounts of nanoparticles. However, the nanoparticle size is limited to nanoparticles larger than 50 nm and a wide size distribution is obtained. In the "bottom-up" approach, nanoparticles are constructed from small building blocks such as atoms or molecules, including for example chemical reduction methods,[52,53] laser-induced-assembly[54] or colloidal aggregation.[55] In this case, the size of the nanoparticle and the agglomeration behavior are effectively controlled by stabilizing the reactive surface with ligands. Moreover, by the right choice of ligands the form and shape of the nanoparticles can be controlled in various geometries such as spheres, rods, cubes, triangles, prisms, cages and stars.[56–60]

In 1951 the synthesis of gold nanoparticles (AuNP) was pioneered by Turkevich:[61] Tetrachloroauric acid was reduced with sodium citrate as a reducing and stabilizing agent to obtain AuNPs in aqueous solution. Over time, syntheses were improved by the variation of the solvent and ligands to obtain different types of AuNPs.[62] Moreover, the results were transferred to synthesize other plasmonic nanoparticles such as silver nanoparticles (AgNP).[63] Interestingly, the general mechanism of nanoparticle formation remains the same and can be subdivided into three steps: nucleation, nanoparticle growth and formation of size distribution.[64] The nucleation describes the formation of nuclei that can act as template for crystal growth. Considering the thermodynamics of the particle formation, the total free energy ΔG is defined as the sum of the surface free energy γ and the free energy of the bulk crystal ΔG_V. For a spherical particle of radius r, the total free energy is described according to equation 2.1.:[64]

$$\Delta G = 4\pi r^2 \gamma + \frac{4}{3}\pi r^3 \Delta G_\mathrm{V} \quad (2.1)$$

The bulk crystal free energy ΔG_V in turn, depends on Boltzmann's constant k_B, the temperature T, the supersaturation of the solution S and its molar volume v:

$$\Delta G_v = \frac{-k_B T \ln(S)}{v} \quad (2.2)$$

A critical supersaturation is needed to initiate the particle formation. Since the surface energy always adopts a positive value and the crystal energy a negative value, it is possible to find a maximum free energy (Figure 2.1A). By differentiating ΔG with respect to the particle radius and setting it to 0, the critical radius r_{crit} with the maximum free energy is obtained:

$$r_\mathrm{crit} = -\frac{2\gamma}{\Delta G_\mathrm{V}} \quad (2.3)$$

$$\Delta G_\mathrm{crit} = \frac{4}{3}\pi \gamma r_\mathrm{crit}^2 \quad (2.4)$$

The critical radius represents the minimum size that a particle needs to reach to avoid redissolving in solution, whereas the critical free energy is equivalent to the activation energy for nuclei/particle formation. After the nuclei are formed, the nuclei grow to form the nanoparticle. The growth of nanoparticles depends on the monomer diffusion to the surface and the reaction of the monomers on the surface. The first mechanism for nucleation and growth of nanoparticle was described by the LaMer burst nucleation mechanism (Figure 2.1B).[65,66]

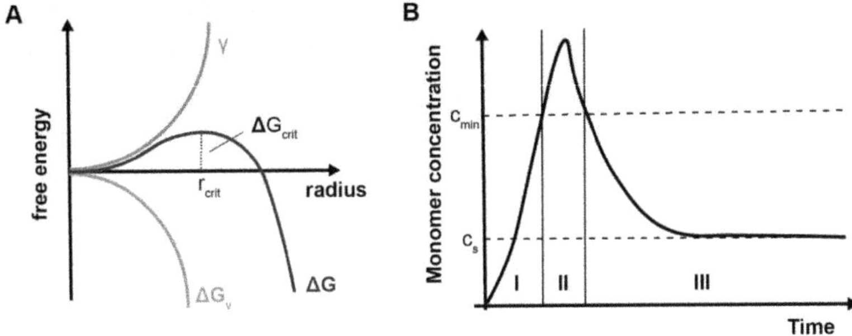

Figure 2.1: Nanoparticle nucleation and growth. (A) Free energy diagram for nucleation. (B) Classic LaMer mechanism for nanoparticle growth.[66]

Within the LaMer mechanism, the reaction progress is subdivided into three phases. Initially, (I) the precursor is converted and the concentration of free atoms increases until the saturation concentration (c_s) is exceeded forming a supersaturated solution. When the so-called critical supersaturation concentration (c_{min}) is reached (II), burst nucleation events occur. Nuclei are simultaneously formed and the concentration of free atoms is drastically reduced. After lowering c_{min}, no further nucleation occurs and (III) the nuclei growth to nanoparticle starts until the concentration of free atoms reaches c_s again. The formation of size distribution happens after the consumption of available free atoms. In general, two different types of mechanism are described. Smaller particles are dissolved due to their unfavorable surface to bulk energy ratio and allow larger nanoparticles to grow even more (Ostwald ripening).[67] Coalescence of particles occurs when two particles of arbitrary size merge and form a larger particle with inconsistent crystal domains.[68] Ideally, particles would grow unhindered and form bulk material in the end, if enough free atoms were available. Therefore, the particle growth needs to be stopped by deactivating their surface with ligands.[69]

The surface functionalization is crucial for the preparation of nanoparticles for potential applications because it controls their interaction with the environment.[70] Hence, the solubility,[71] colloidal stability,[72] or binding affinity[73] are affected. In general, the choice of ligands is predetermined by the solvent used for the nanoparticle synthesis. In apolar organic solvents, the nanoparticle surface is covered by hydrophobic ligands to prevent the aggregation of the nanoparticles. Here, ligand molecules with long or bulky hydrocarbon chains are used to ensure steric repulsion between nanoparticles. In polar or aqueous solutions, electrostatic repulsion plays a crucial role for the nanoparticle stabilization. In contrast to the steric repulsion, the electrostatic repulsion is highly affected by external influences. High salt concentration can shield the ligand electric field so that nanoparticles can come closer to each other until the attractive forces cause the particle agglomeration.[74] Moreover, the surface charge depends on the pH value, and thus influences the nanoparticle interaction between each other. Consequently, the steric stabilization is much more resistant than electrostatic stabilization. One requirement for the ligand molecule is that it binds to the particle surface by some attractive interactions such as chemisorption, electrostatic attraction, or hydrophobic interaction. Usually, binding to the surface is provided by a head group of the ligand molecule for example thiol,[75] amine[76] or phosphine[77] atoms. However, the inorganic surface and the electron donating-group may undergo a dynamic binding and desorption process. These dynamics enable the possibility to replace the attached ligands with others. Since many nanoparticle syntheses are performed in organic solvents, even a phase transfer is possible with a ligand exchange reaction to obtain water-soluble nanoparticles. Thiol ligands are commonly used for ligand exchange reactions because the thiol group possess the highest binding affinity to noble metal surfaces.[78] For example, AuNPs stabilized with phosphine ligands[79] as well as amine ligands[80] were transferred to aqueous solution by replacing the ligand shell with thiol-based ligands. Moreover, a mixture of different ligand molecules were employed to obtain so-called Janus nanoparticles.[81] The ligand exchange in the same solution can be challenging especially for polar ligands. An exchange from a positively charged ligand to a negatively charged ligand or vice versa can lead to unwanted aggregation of nanoparticles because of the detrimental electrostatic interaction between the two ligands. A two-step phase transfer including a phase-transfer to uncharged hydrophobic ligands can simplify the surface functionalization.[82]

2.1.2 Plasmonic nanoparticle properties and applications

Plasmonic nanoparticles can be distinguished from other nanoparticles such as metal oxide nanoparticle or carbon-based nanoparticles by their unique optical properties.[83] Since the antiquity, optical properties of colloidal gold were incorporated in materials for example in ceramics as ruby glass. However, in 1857 Faraday was the first to attribute these fascinating behavior to the optical properties of colloidal gold.[84] He reported the formation of a deep red solution of colloidal gold by reduction of chloroaurate and observed reversible color changes of thin films upon mechanical compression. The optical properties of plasmonic nanoparticles originates form the local surface plasmon resonance (LSPR). Plasmon resonance is an optical phenomenon arising from collective oscillation of the electron cloud in a metal when conduction electrons are disturbed from their equilibrium position (Figure 2.2).[62]

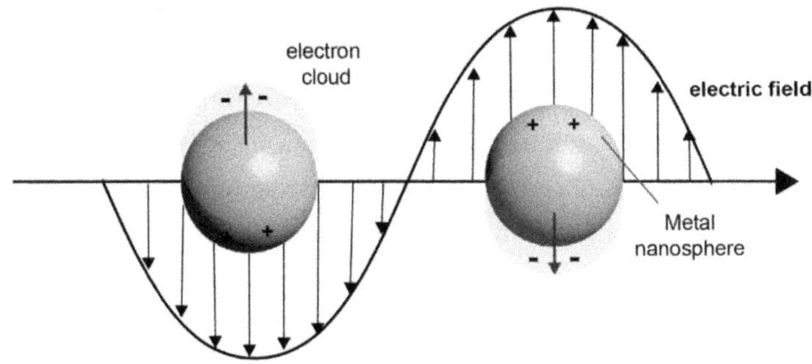

Figure 2.2: Schematic illustration of the local surface plasmon resonance. [85]

In detail, an incoming electromagnetic wave (light) will interact with the particle and immediately induce the conduction electrons to move from the equilibrium position. The small size of the particle prevents the free movement of the charge, as it would be possible in the bulk material. Thus, the electrons accumulate on the particle boundary. Coulomb attraction between the positively charged ions in the metal lattice and the negatively charged electrons at the particle boundary forces the electrons back to the equilibrium position, leading to an oscillation of the free electrons. When the oscillation frequency is approximately the same as the incident light, a resonance condition is reached, resulting in one plasmon resonance peak. Importantly, the position of the plasmon resonance peak is affected by the type of metal,[86] surrounding dielectric medium[87,88] and the electromagnetic coupling of the nanoparticles.[89,90] Any variation in the shape or size of the particle can alter the surface polarization.[91] The interaction of the electromagnetic wave with the nanoparticle can be understood by solving Maxwell's equation. The Mie theory enables the calculation of extinction spectra but it is limited to spherical particles of diverse sizes.[92]

The great interest in plasmonic nanoparticles is driven by their potential applications. Plasmonic nanoparticles can compress light signals down to the nanoscale. They can form the basis of novel nanoscopic optical devices for light concentration and guiding in plasmon optics. For example, the change in LSPR peak position and intensity can be used for optical sensing.[93] Due to the dependency of the LSPR peak position on the refractive index, molecular binding events can be monitored. For example, the detection of human immunodeficiency virus (HIV) was possible by the alteration of refractive index caused by the adsorption on the nanoparticle surface.[94] Moreover, the same concept can be utilized to perform immunoassays without additional labelling material.[95] As an alternative approach, the plasmon coupling between single nanoparticles can be exploited to report conformational changes and intramolecular distances of single biomolecules similar to Förster resonance energy transfer (FRET).[96] Furthermore, molecular rulers based on fluorophores suffer from low and fluctuating signal intensities, limited observation time due to photobleaching and an upper distance limit of approximately 10 nm. In contrast, plasmonic nanoparticle-based molecular rulers do not bleach and larger distances can be measured. Another feature of plasmonic nanoparticles is the enhancement of the local electromagnetic field by the LSPR. This effect is utilized in surface enhanced Raman spectroscopy (SERS) to improve Raman scattering cross section. Enhancement factors of 10^4 to 10^{11} are obtained.[97] More importantly, plasmonic nanoparticles can be utilized for the transport of optical signals in plasmonic waveguides due to the ability to manipulate electromagnetic signals below the diffraction limit of light.[98] A significant prerequisite for this application is the close and precise positioning of the plasmonic nanoparticles to enable strong near-field coupling between adjacent nanoparticles.[99] In this way, highly ordered structures of plasmonic nanoparticles can guide electromagnetic energy by converting optical modes into non-radiating surface plasmons.[100] Interestingly, this property can be exploited for optical devices in computing and informational technology.[101] According to Moore's law the development of fast electronic devices increased. However, the current electrical interconnection based on silicon-derived components reaches their limit. Electrons as information carrier cause signal delay and thermal issues, preventing significant increase in processor speed. In contrast, optical devices possess a large data-carrying capacity (bandwidth) by using light as efficient information carrier.[102] On the one hand, an increased information density is achieved using several frequencies simultaneously.[103] On the other hand, the same scaling of the devices as in electronics is hampered by the fundamental laws of diffraction, complicating the interconnection with other electronic devices and thus the application.[101] Although optical devices are used for data transmission for example in optical fiber cables,[104] the shift in computing from conventional devices based on electrons as information carrier to optical devices remains challenging.

The common feature of all optical devices is the precise organization of the plasmonic nanoparticles into highly ordered materials to enable strong near-field coupling between neighboring nanoparticles. In the next chapter, general methods for the formation of nanoparticle superlattices are presented.

2.2 Formation of nanoparticle superlattices

For the assembly of nanoparticles, single particle manipulations are not feasible since typical lithographic techniques cannot manipulate small nanoparticles for the creation of complex three-dimensional arrays.[105] In contrast, self-assembly of nanoparticles into colloidal, crystalline systems is seen as a promising strategy.[12] Self-assembly can yield structures with unrivalled precision in positioning of individual particles and is able to simplify the generation of metamaterials considerably.[13] In the following section, current approaches for self-assembly of nanoscale building blocks are highlighted.

2.2.1 Self-assembly with organic ligands

Self-assembly describes the spontaneous organization of building blocks into highly ordered structures by direct interactions or indirectly using a template. The obtained structures are modulated by carefully choosing and constructing the building block. In general, ligand-stabilized nanoparticles can be considered as soft spheres with an effective hard sphere assigned to the inorganic core. The inorganic core provides interesting properties for the material, and the ligand shell is responsible for the interaction between building blocks. These interactions can be altered chemically (surface functionalization) or with external stimuli (pH, temperature, solvent polarity). This way, the structure can be controlled by manipulation of the guiding forces for the assembly such as hydrophobic interactions, hydrogen bonding or molecular dipole-interactions.[106]

Spherical nanoparticles with a narrow monomodal size distribution can be easily ordered directly after the synthesis, when the solvent is allowed to evaporate in a controlled fashion. Monolayer films are formed in cubic close-packed or hexagonal close-packed structures.[107,108] Moreover, nanoparticles with a bimodal size distribution can organize into two-dimensional arrays.[109] Due to the effective hard sphere of nanoparticles, the formation of binary structures should proceed similar to conventional inorganic salt. In inorganic salts, the structure of two different atoms with atomic radii r_A and r_B depends mainly on the ratio of the radii (r_A/r_B) and the ratio of atoms (n_A/n_B).[110] Several calculations were performed to identify preliminary ranges of size and number ratio to lead to specific structures.[111,112] For example, when $n_A/n_B = 1$ and $r_A/r_B = 1$, an AB structure analogous to CsCl is expected. Alternatively, when $n_A/n_B = 1$ and $0.27 \leq r_A/r_B \leq 0.425$, an AB structure analogous to NaCl is expected. Kiely *et al.* showed that a mixture of differently sized AuNP and AgNP forms monolayer arrays with an AB_2 structure.[113] Importantly, the formed structure agrees with the expected AB_2 structure similar to an AlB_2 alloy structure. The distance between the inorganic cores is predetermined by the ligand shell thickness. In a next step, the formation of binary structures has been extended from monolayer arrays to binary nanoparticle crystals. For instance, the use of specific size ratios between PbSe semiconductor quantum dots and Fe_2O_3 magnetic nanoparticles directs the assembly to AB_{13} or AB_2 structures.[114] In the AB_{13} structure, the larger Fe_2O_3 nanoparticles form a cubic lattice with an icosahedron of smaller PbSe quantum dots in the middle isostructural to $NaZn_{13}$. In contrast, the AB_2 structure is composed of a hexagonal unit cell isostructural to AlB_2. Single domains of AB_{13} and AB_2 were formed in the size range from 0.16 μm^2 to 2 μm^2 separated by pure nanoparticle superlattices. Furthermore, the assembly of iron and gold nanoparticles to a three dimensional AB nanoparticle superlattice was performed.[115] Again, domains of pure and binary nanoparticle superlattices appear side by side. In all these materials, the control over the particle packing is strongly limited because of the isotropic ligand shell.

In an extension of this approach, a colloidal model system with oppositely charged nanoparticles is generated to form binary ionic crystals.[116] Grzybowski *et al.* showed the formation of a binary nanoparticle superlattice with a crystal size up to 3 µm using oppositely charged AuNP and AgNP.[117] The surface is charged by functionalization with alkane thiols with either carboxy or tertiary amine groups. After assembly, a diamond-like lattice was formed. Each NP is surrounded by four other nanoparticles, similar to the ZnS structure, except here the particles have the same radii. Interestingly, highly ordered crystals were only formed from nanoparticles with a similar size distribution, because smaller and larger nanoparticles are entropically disfavored.

Finally, Shevchenko *et al.* showed the remarkable structural variety by using different oppositely charged nanoparticles for the formation of binary superlattices (Figure 2.3).[13] The charge of the nanoparticles was altered by adding surfactants. In total, 15 different lattices were observed including ones which are not isostructural with specific intermetallic compounds. Superlattices with AB, AB_2, AB_3, AB_4, AB_5, AB_6 and AB_{13} with cubic, hexagonal, tetragonal and orthorhombic symmetries have been identified. Importantly, 11 different structures were obtained from the same batch of 6.2 nm PbSe and 3 nm Pd nanoparticles by altering the surface charge of the nanoparticles. Furthermore, different structures were generated by combination of triangular LaF_3 plates and spherical AuNPs or PbSe nanoparticles.

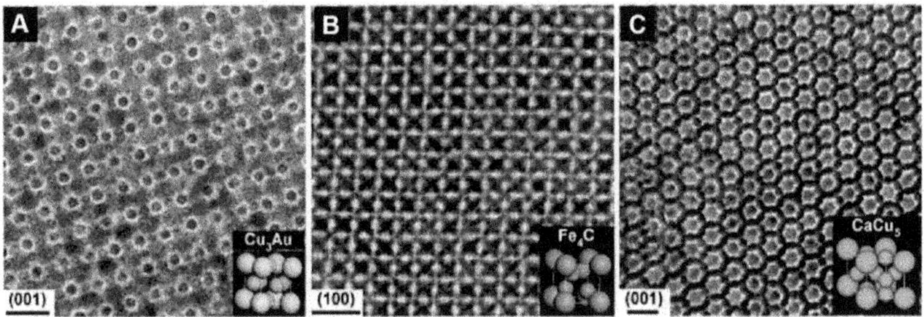

Figure 2.3: Binary nanoparticle superlattices. Nanoparticle superlattices are composed of (A) 7.2 nm PbSe and 4.2 nm Ag, (B) 6.2 nm PbSe and 3.0 nm Pd and (C) 7.2 nm PbSe and 4.2 nm Ag. Figure reprinted from reference 13 with permission from Springer Nature, copyright 2006.

Currently, the structural diversity of the binary superlattices is difficult to understand because of the complex interplay of multiple factors such as entropy, different type of interactions (Coulombic, van der Waals, dipole-dipole) or type of nanoparticle (size, composition, shape). Moreover, only crystals with small domain sizes are obtained due to the polydispersity of the nanoparticles.

2.2.2 Assembly with DNA molecules

The crucial influence of the nanoparticle and its ligand shell on the structure makes development of new programmable building blocks essential. Perfect candidates are biomolecules because of their structural and functional diversity. In the last 15 years, DNA molecules were frequently used because a precise control of the strength, length and interparticle specificity of bonding interactions is possible.[118,119] The general idea is the functionalization of the nanoparticle surface with synthetic single-stranded DNA (ssDNA) molecules. By introduction of a linker, nanoparticles with a free single-strand end can be selectively programmed to induced assembly.[120] The groups of GANG and MIRKIN showed independently the DNA guided assembly of AuNP into different crystalline states (Figure 2.4).[20,21]

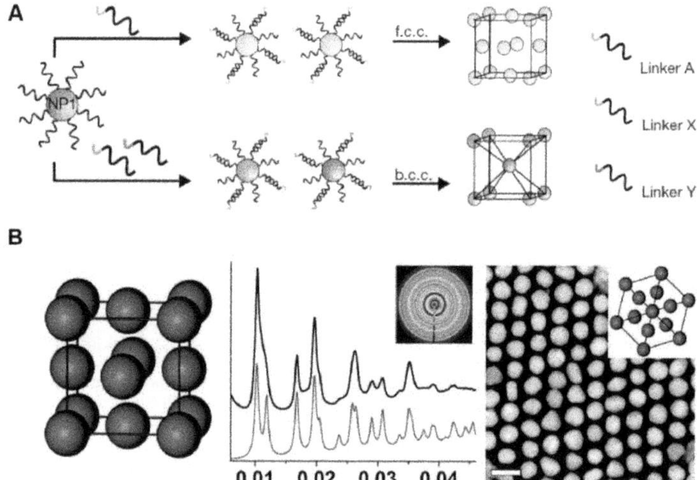

Figure 2.4: Nanoparticle assembly with DNA linker. (A) General strategy for the assembly of nanoparticles with DNA linker strands into a highly ordered structure. Figure adopted from reference 21 with permission of Springer Nature, copyright 2008. (B) Assembly of nanoparticles in a face-centered cubic crystal structure. SAXS data and TEM image confirm the highly ordered structure. Figure adopted from reference 121 with permission of The American Association for the Advancement of Science, copyright 2011.

The GANG group exploited the interaction between complementary DNA molecules attached to the AuNP surface for the assembly. The formation of the superlattice was investigated for different lengths of ssDNA molecules while keeping the same linker DNA. AuNPs with short ssDNA molecules tend to form amorphous precipitate. However, AuNPs with long ssDNA molecules crystallize spontaneously into highly ordered structures. A body-centered cubic lattice isostructural to CsCl was formed. SAXS measurements of the crystals revealed a long-range order of approximately 0.5 µm. Due to the DNA molecules, a flexible nanoparticle superlattice was obtained. By varying the temperature, the unit cell length was altered and reversible dis- and reassembly of the assembly was performed during heating and cooling cycles. At the same time, the MIRKIN group showed the influence of the DNA sequence choice for the assembly structure. By performing the assembly with self-complementary ssDNA molecules, a face-centered cubic lattice was formed. Each nanoparticle possesses twelve neighboring nanoparticles. The change to non-self-complementary ssDNA molecules guided the lattice formation to a body-centered cubic lattice. Here, each nanoparticle possesses eight neighboring nanoparticles. Compared to the GANG group, slightly larger crystals (~2.1 µm) were obtained. Shorter ssDNA molecules or the combination of short and long ssDNA resulted in the body-centered cubic lattice with shorter unit cell axis. Importantly, high monodisperse nanoparticles are required for the assembly. Six design rules were defined for the assembly of nanoparticles with DNA molecules. By adjusting the attached DNA molecules on the nanoparticle surface, nine different distinct crystal structures were obtained with control over lattice parameters.[121] This variability includes AB, AB_2, AB_3, AB_6 and hexagonal close-packed structures. In later work, a third nanoparticle component was introduced at predetermined sites within a preformed binary lattice by topotactic intercalation.[23] In detail, a known binary superlattice was formed and the third component was added to form a distinct ternary structure. In total, five new structures were obtained including three structures, which have no equivalence in atomic or molecular crystals. Moreover, a switch between a two-component and three-component structure was readily performed by changing the temperature.

To reduce the influence of the nanoparticle type on the lattice formation, individual DNA constructs with programmable architectures were developed as template for the nanoparticle organization. For the first generation of building blocks, the Holliday junction – a branched conformation of DNA – was the basis to construct so-called DNA tiles.[122] Linking between different building block was enabled by introduction of unbound regions of ssDNA at the loose ends of the DNA tiles (sticky ends). The first immobile junctions were a tetrameric and three-armed branched junction.[123] Complex structures like cubes[124] or truncated octahedrons[125] were built up by combination of these two building blocks. Moreover, more branched junctions were developed to create new architectures.[126,127] However, an increased degree of branching reduces the stability of the construction. As an extension, DNA double-crossover (DX) molecules were introduced as new motif to increase the mechanical stability.[128] In this motif, DNA molecules with two crossover sites between helical domains were exploited. In total, five unique structural arrangements were designed as building block for the assembly in 2D structures.[123-127] For example, two 3D DX triangles were designed and combined to produce a rhombic lattice arrangement.[134] Thiolated ssDNA was attached to the AuNP surface and combined with the 3D DX triangle to obtain 3D DX triangle-Au conjugates. Importantly, by mixing the 3D DX triangle-Au complex with the complementary 3D triangle, the rhombic lattice arrangement was formed again, with AuNP ordered in a highly structured arrangement. The robustness of the system was demonstrated by successful organization of the nanoparticle superlattice by combining either one 3D DX triangle loaded with AuNP or both. Moreover, the assembly of two different sized AuNP was possible. Nevertheless, the lack of flexibility in DNA tiles limits the number and complexity of possible structures.

The construction of DNA based architectures was further improved by the so-called DNA origami technique.[135] Comparable to the art of paper folding, a long ssDNA (scaffold sequence) is fixed with shorter strands (staple sequence) at certain positions. Interactions between scaffold sequence and staple sequence lead to the self-assembly of the designed shape. With this method, highly symmetric shapes like spheres, rectangles, stars or triangles[136] were formed as well as asymmetric shapes such as dolphin-like conformations.[137] Moreover, the formation of 3D structures such as a honeycomb lattice[138] or cage-like polyhedral geometries[139] were feasible. Especially, the formation of cage-like structures as templates for the nanoparticle assembly have gained a lot of interest. Gang *et al.* showed that the same type of nanoparticle can be assembled with different frame geometries into different crystal lattices (Figure 2.5A).[22] Five different polyhedral cages with cube, octahedron, elongated square bipyramid, prism and triangular bipyramid geometries were designed. Coordination of the nanoparticle was performed at the vertices of the cage-like structures. Interestingly, the geometry of the frame strongly influences the formed lattice type due to the binding of nanoparticles at the vertices of the frame. For example, the assembly with octahedron shaped cages results in a face-centered cubic structure. Instead, a body-centered tetragonal structure was obtained with elongated square bipyramid shaped cages. However, the nanoparticle size strongly affects the lattice formation. On the one hand, if the nanoparticle is significantly larger than the DNA-origami polyhedral frame, the lattice topology is disrupted because of many attached frames on the nanoparticle surface. On the other hand, if the nanoparticle is too small, the hybridization of multiple frames for each particle is hindered because of steric repulsion between the frames.

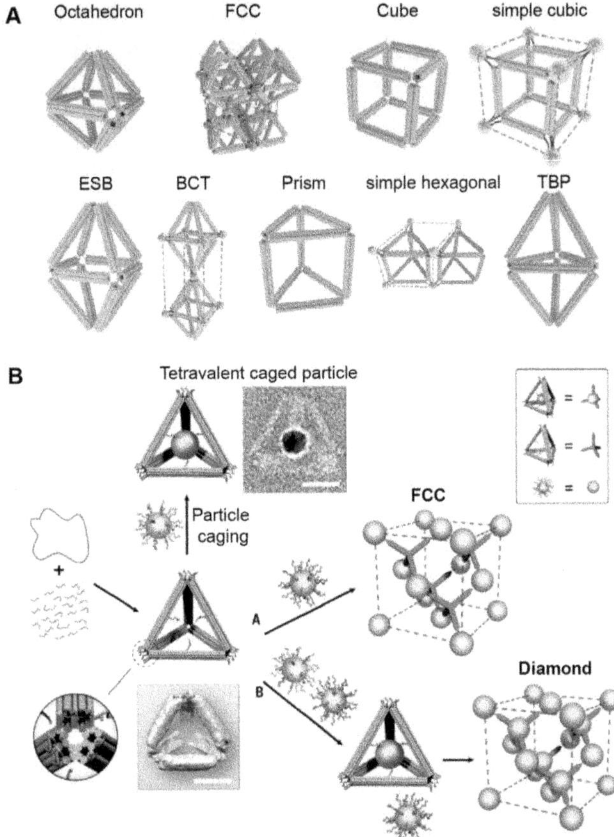

Figure 2.5: Nanoparticle assembly by DNA origami. (A) Different DNA origami cage-like architectures are used to assemble AuNP by linking the vertices of the cages into highly ordered structures. The structure depends on the DNA origami frame. Figure adapted from reference 22 with permission of Springer Nature, copyright 2016. (B) Diamond-like structure achieved by assembly of uncaged and caged AuNPs with tetrahedral DNA origami cages. Figure adapted from reference 140 with permission of The American Association for the Advancement of Science, copyright 2016.

Besides the incorporation of nanoparticle at the vertices of the cage-like building blocks, the DNA-functionalized nanoparticles can be incorporated inside the cage (Figure 2.5B).[140] Tetrahedral cages as building blocks form together with AuNPs a face-centered structure. However, by caging AuNPs inside the tetrahedral cage and co-assembly with uncaged nanoparticles, a diamond like lattice was obtained. The formation of a diamond-type lattice is considered as highly challenging because of the low space filling of the lattice type. Interestingly, an exchange of the nanoparticles to smaller ones led to a zinc blende structure with only uncaged nanoparticles and to another previously unknown lattice with both types of nanoparticles. Incorporation of nanoparticles inside the cage-like structure remains interesting to minimize the influence of the nanoparticle size for the assembly. Wang *et al.* showed the incorporation of DNA functionalized AuNPs inside an octahedral DNA-origami cage.[141] Assembly of the octahedral frames was performed by linking the corners of the octahedral frames with DNA. Importantly, the assembly behavior of the octahedral cages was controlled by the number and positions of the DNA linker. For example, if the four corners in the middle plane are encoded with DNA linkers, highly ordered 2D square planes are formed. Additional DNA linkers at the top or bottom position guide the assembly to a 3D structure but with small lateral dimensions.

Despite all the advantages of DNA molecules for the assembly of nanoparticles, namely the assembly into a variety of different structures and assembly types, still several limitations exist. The overall domain sizes remain rather small (i.e. in the low micrometer range). The previously discussed approaches mostly rely on nanoparticles as the primary building block, which cannot be synthesized with atomic precision because of the intrinsic disorder at the atomic scale.[142] As a consequence, the material has a low range order due to the polydispersity of the nanoparticles. Furthermore, a large size distribution prevents crystal formation. However, the construction of mesoscale materials has been identified as a major challenge in material research.[143] In addition, the distances between the nanoparticles in the lattice are usually 20-50 nm. The large distances between the building blocks hamper strong interactions between neighboring nanoparticles, therefore making these lattices not ideal as in optical materials. Moreover, the crystals possess a large solvent-content and thus are not stable outside their growth medium. In conclusion, an alternative approach is required, which on the one hand, will enable short interparticle distances for a strong interaction between the building blocks and on the other hand, will assemble into a crystalline material with a high-long range order. Proteins, another type of biomolecule, are another interesting building block for the construction of highly structured materials. In nature, proteins are produced in a programmed synthesis with each chain having the same composition guaranteeing a narrow size distribution. Moreover, the self-assembly of proteins into higher multimeric complexes such as protein container enables an easy access to cage-like structures for the incorporation of nanoparticles.

2.3 Protein containers as a new building block

Protein containers are a special class of proteins, which form by precise assembly of protein subunits into cage-like architectures.[144,145] Usually two classes of protein containers are defined based on the origin: virus-like particles and non-viral protein containers. Viruses are nanocontainers that store their genetic information for replication within their nanocage. Importantly, the metastable structure of viruses withstands extreme environments but at the same time is able to release their nucleic acid cargo in the appropriate cellular environment. Moreover, the high symmetry of the nanocontainer enables the construction from one or a few different types of subunits. Thus, only a small amount of genetic information needs to be stored in the viral cage. Protein containers with viral origin, but without viral activity and genetic information stored in the viral cage, are referred to as virus-like particles. On the one hand, non-viral protein containers are nanocontainers, which are exploited as a storage nanocompartment or as reaction chamber. Common feature of all protein containers is the assembly from a limited number of subunits to form robust nanostructures. Often, spherical structures are formed, but also rods, rings and more complex structures are possible.[146,147] Generally, special focus is placed on spherically cages based on icosahedral, cubic or tetrahedral symmetries. The spherical geometry of the protein containers provides precisely defined exterior and interior surfaces, resulting in three chemically distinct interfaces: the interior surface, the exterior surface, and the interface between subunits. Importantly, the manipulation and redesign of the subunits by chemical and/or genetic alteration enable the generation of novel functionality. This way, container properties are affected such as encapsulation of cargo,[148] targeting of the protein container[149] or assembly into hierarchical structures.[40] The cavity of the protein container is accessible via the container pores or by reversible dis- and reassembly of the cage through extern stimuli. The defined interactions between protein subunits can be disturbed by changes in pH[150] or adding salt[151] to disassemble the protein cage into their subunits. Reassembly of the cage is performed by adjusting the original interactions between the protein subunits. The pores can function as selective gates to prevent larger molecules from diffusing into the cavity.[152,153] Moreover, more specific access control is enabled due to hydrophobic or electrostatic character of the pores to distinguish between molecules with a distinct polarity or charge.[154] Despite all the possibilities to modify the protein container, no single protein container has all requisite properties regarding size, chemical/thermal stability and functionality. Over the years, researchers have investigated a library of several spherical natural nanocontainers. A selection of several protein containers is shown in Figure 2.6.

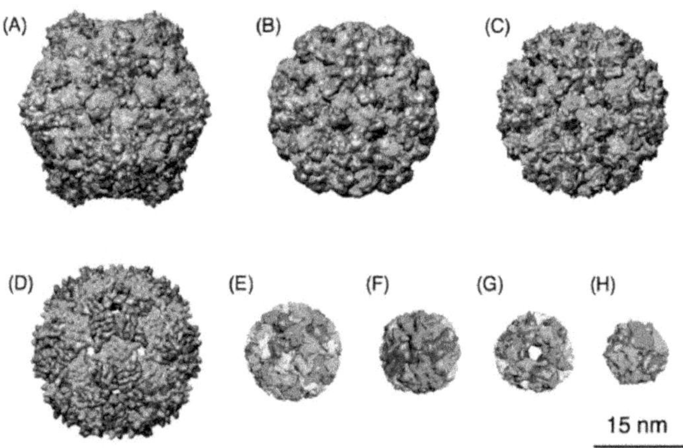

Figure 2.6: Examples of natural protein containers. (A) Cowpea mosaic virus (31 nm), (B) Brome mosaic virus (28 nm), (C) Cowpea chlorotic mottle virus (28 nm), (D) MS2 bacteriophage (27 nm), (E) lumazine synthase (15 nm), (F) ferritin (12 nm), (G) small heat shock protein (12 nm) and (H) DNA binding protein from starved cells (9 nm). Figure reprinted from reference 144 with permission of John Wiley and Sons, copyright 2007.

In addition to naturally occurring protein containers, novel artificial containers were designed. Typically, native oligomeric proteins were used and assembled by computationally designing binary protein-protein interactions,[155–157] disulfide bond formation[158] or metal coordination.[159,160] However, these artificial protein containers are highly porous. These drawbacks were tackled utilizing chemically orthogonal coordination motifs with metal ions, leading to protein containers with small pore sizes.[161] Moreover, different coordination sites for metal ions were introduced by genetically modification of the monomers to alter the structure from a dodecameric cage into a hexameric cage.

2.3.1 The ferritin protein container

Ferritin is an iron storage protein for biomineralization,[25] which can be found in almost all organism. It is composed of 24 structurally identical subunits forming a rhombic dodecahedron cage, crystallizing for example in a *F*432 space group (Figure 2.7A).[162] The cage-like structure possesses an outer diameter of 12 nm and an inner diameter of 6-8 nm. For the monomer, the structural motif consists of five α-helices, which form a four-helix bundle with a small left handed twist (E helix) orientated 60° to the four-helix bundle axis (Figure 2.7B).[163]

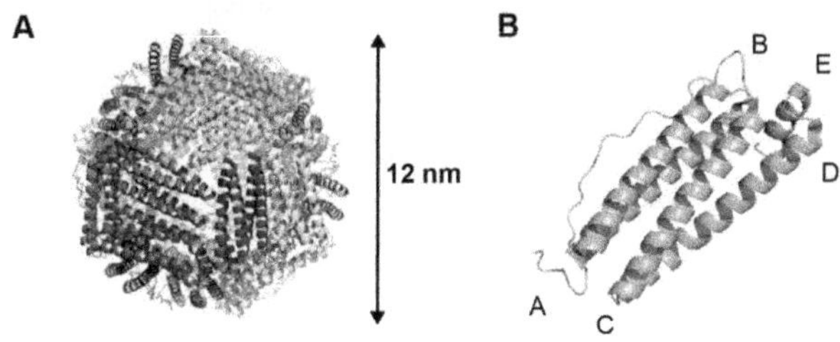

Figure 2.7: Ferritin protein container. (A) Cartoon representation of the ferritin protein container (PDB: 2cei). View along the threefold axis. (B) Close-up of the protein subunit of the ferritin protein container. Helices are annotated.

Moreover, eight hydrophilic funnel-like channels are located at the three-fold axis and six channels at the four-fold axis. Ferritin is a heteropolymer and consists of two classes of subunits that are nearly structural identical: heavy chain (H-chain) and light chain (L-chain) subunits.[164] The H-chain subunit has an molecular mass of 21 kDa and the L-chain subunit of 19 kDa. Beside the difference in mass, the general composition of both subunits differs. On the one hand, the H-chain subunit possesses a di-iron binding site (ferroxidase site), which binds and oxidizes ferrous ions.[165] On the other hand, the L-chain subunit possesses only a favorable nucleation site, but no catalytic ferroxidase site.[166] Hence the ratio of the subunits determines the function of the ferritin container. H-chain rich ferritin performs fast iron metabolism, but L-chain rich ferritin is suitable for long term iron storage.[167] The ratio of both subunits varies between organisms. For instance, the most frequently used ferritin protein container in research, commercially available horse spleen ferritin, is composed of approximately 90 % of L-chain subunits. In general, iron oxide formation in the ferritin container cavity can be described as follows: iron ions entry the cavity through the three-fold channels. The electrostatic field gradient directs cations to the channel entrance and guides them inside the cavity.[167] At the ferroxidase center of the H-chain subunit, two Fe(II) ions are catalytically oxidized by consumption of oxygen and the intermediate product H_2O_2, forming the iron oxide nanoparticle core according to equation 2.5:

$$4\,Fe^{2+} + O_2 + 6\,H_2O \rightarrow 4\,Fe(O)OH + 8\,H^+ \quad (2.5)$$

The resulting iron oxide nanoparticle can be described as hydrated ferric oxide (ferrihydrite) with the approximately stoichiometry of 5 $Fe_2O_3 \cdot 9\,H_2O$ and storing up to 4500 Fe(II) ions per container.[166] The inorganic core of the ferritin protein container can be removed without altering the structure, leaving a colorless protein called "apoferritin". Ferritin and apoferritin show a high stability up to 85°C and tolerate reasonably high levels of urea and guanidinium chloride.[168,169]

In recent years the ferritin container was severely exploited as a reaction chamber for the synthesis of a large variety of different nanoparticles.[25] Beside nanoparticle synthesis, different organic metal complexes were inserted through the pores inside the cavity to perform catalytic reactions inside the protein container cavity.[153,170,171] The large protein container cavity enables the incorporation of a vast amount of metal organic complexes and even of two in the same cavity.[172] Moreover, in case the pore size prevents the incorporation of a large molecule, the dis- and reassembly behavior of the protein container enables the incorporation by this route. With this strategy, the loading of ferritin with antibiotics for drug-delivery applications was performed.[173,174] Nevertheless, the utilization of the ferritin container is not restricted to the native ferritin protein container. The well-known structure to atomic resolution enables the modification of the structure, thus chemically or genetically customizing the protein container to the desired function.[25] Importantly, depending on the application, one specific interface can be altered (Figure 2.8).

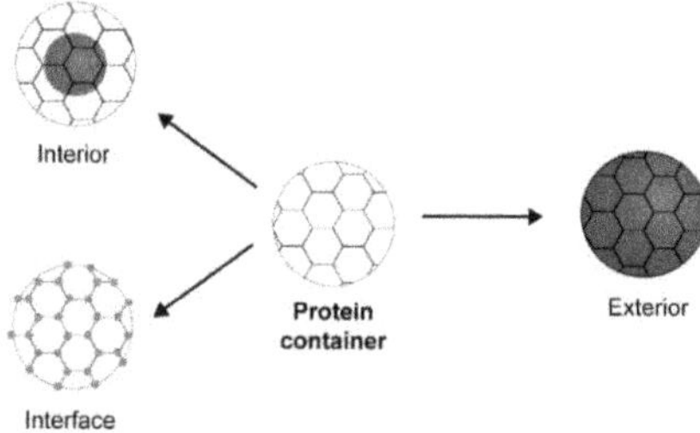

Figure 2.8: Interfaces in the protein container. Schematic illustration of the three interfaces in the protein container for genetical and chemical modification. Figure adapted from reference 144 with permission of John Wiley and Sons, copyright 2007.

The genetical modification of the inner surface can affect the properties of the cavity significantly. By mutation of amino acids, new metal bindings sites or hydrophobic pockets can be introduced to alter the nanoparticle synthesis or the adsorption of molecules.[29,175] Moreover, the alteration of the surrounding of specific binding sites can enhance the catalytic activity of incorporated molecules by adjusting a beneficial coordination geometry.[176,177] By modification of the interface between two protein subunits, new assembly behavior can be created. For example, the selective deletion of a small number of amino acid residues strongly affects the assembly of the protein subunits to the protein container. The assembly of the native 24-mer ferritin with a diameter of 12 nm can be guided to a non-native 48-mer variant of ferritin with a diameter of 17 nm.[178] Interestingly, by cleavage of the last 23 amino acids at the carboxyl terminal, the disassembly of the protein cage was possible at milder conditions (pH 4) with successful reassembly at pH 7.5. In this way, a higher encapsulation efficiency of the molecule curcumin was achieved compared to the native ferritin protein container.[179] Last but not least, alteration of the outer surface affects the interaction of the protein container with the environment or each other. Chemical modification of the protein container outer surface can influence the solubility of the protein. Usually, proteins are dissolved in aqueous solution and precipitate quickly with addition of small amounts of organic solvents. A transfer to organic solvents would be advantageous to increase the field of applications. The solubility in organic solvents can be enhanced by derivatization of the protein shell with long-alkyl chains.[180,181] Towards creating defined nanostructures, the assembly behavior of the ferritin protein containers to form hierarchically ordered assemblies was adjusted by mutation of amino acids on the outer surface to positively charged amino acids. Thus, the incorporation of positively surface-charged ferritin containers inside larger capsid-forming enzyme lumazine synthase from *Aquifex aeolicus* was performed to construct Matryoshka-type cage within cage structures.[148]

Since the assembly of the protein containers is the crucial step for the construction of templates for the nanoparticle assembly, the next chapter 2.3.2 will focus on the assembly and tuning of the protein crystal matrix. At the end, the latest results of the nanoparticle assembly by a protein matrix will be presented.

2.3.2 Organization of ferritin into highly ordered structures

The vast field of protein crystallography underscores the importance of protein assembly into ordered structures and arrays.[182] Although protein crystallization is based on a diverse set of principles, experiments and ideas, no comprehensive theory to guide crystallization efforts has been developed to date. Protein crystallization is still largely of empirical nature. Theoretically, protein crystal formation is subdivided in nucleation and growth. Both processes depend on the production of a supersaturated solution through addition of mild precipitating agent (salts, monomers) or by manipulation of various parameters (temperature, ionic strength, pH). The supersaturated solution is a non-equilibrium condition, where disordered proteins form ordered assembly (nuclei) during the nucleation step. By formation of a solid state (crystal growth), the equilibrium is re-established. Importantly, the utilization of practical, easy-to use screening kits and of laboratory robotics significantly increased the success of protein crystallization over the years. Highly automatic pipetting robots accelerate the preparation of enormous quantities of crystallization plates, thus simplifying the search for suitable crystallization conditions. Subsequently, manual modification of the hits by specific alteration of the crystallization parameters is a way to optimize the crystals in size.[183]

In nature, a variety of different structures exists that can be obtained by assembly of the same protein container. For example, horse-spleen ferritin can crystallize in a cubic $F432$ space group or in a tetragonal $P42_12$ space group.[184,185] The main difference between these two structures are the interactions at the crystal contact interfaces. In the cubic structure, only salt bridge type interactions via cadmium ions are formed. On the other hand, both salt bridge type interactions via cadmium and hydrogen bonding interactions are formed in the tetragonal structure. Along these lines, the rational design and modification of the crystal contact interfaces can create modularity in the formation of alternative crystal structures. Tezcan *et al.* showed that such modularity can be obtained by protein-metal-organic frameworks.[186,187] Here, the strategy of metal-organic frameworks was adopted to protein containers (Figure 2.9). A metal coordination site was introduced through the mutation of amino acids on the entrance of the threefold axis pore of the ferritin protein container. Zinc ions were coordinated at this new metal coordination site to assemble the protein containers by linking the ions with an organic linker. The conventional face-centered cubic structure of ferritin was altered to a body-centered cubic structure. By variation of the metal (Zn, Ni, Co) and the organic linkers, cubic or tetragonal structures with different lattice parameters were obtained.

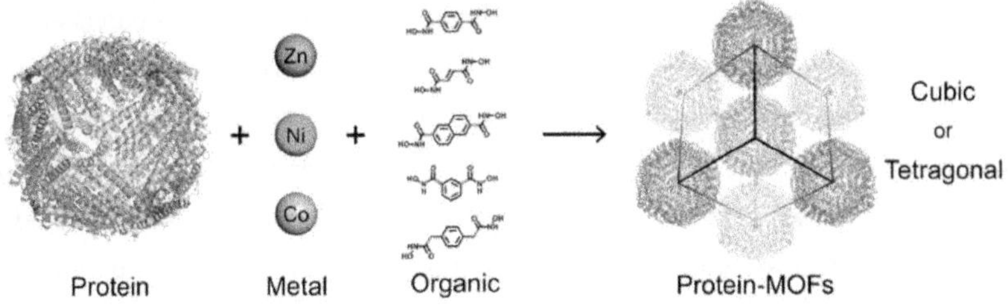

Figure 2.9: Assembly of ferritin into highly ordered structures. Protein-metal-organic frameworks modularity. Figure adopted from reference 187 with permission of the American Chemical society, copyright 2017.

In addition to the metal assisted control, electrostatic assembly provides a second approach to tune the crystal formation. Kostiainen *et al.* showed that the co-crystallization of apoferritin and cationic poly(amidoamine) dendrimers (PAMAM) led to crystals with tunable lattice parameters.[188] By altering the PAMAM size, control over the lattice constant was achieved. Moreover, a change of the ionic strength in the crystallization solution influenced the electrostatic interactions between the building blocks. Depending on the ionic strength, a cubic or hexagonal structure was obtained.

For the assembly of nanoparticles encapsulated inside protein containers into crystalline biohybrid materials two general strategies exist. (I) The nanoparticle is present during the assembly of the protein container into the ordered structure or (II) the protein matrix is formed, and the nanoparticle is synthesized *in situ* inside the protein matrix afterwards. An example for the first strategy was recently established by Beck *et al.* (Figure 2.10A).[40] Here, the protein surface of ferritin was modified to obtain oppositely surface-charged protein containers. Different metal-oxide nanoparticles were synthesized inside the protein container cavity by diffusion of metal precursor and hydrogen peroxide through the protein container pores. Oppositely charged protein containers loaded with nanoparticles were assembled using protein crystallization to form a highly ordered nanoparticle superlattice. The structure of the nanoparticle superlattice is independent of the nanoparticle cargo and only depends on the protein matrix. Thus, the formation of unitary and binary nanoparticle superlattices is feasible. Moreover, lattice parameters are tunable by altering the interaction between the oppositely charged protein containers.[189] Formation of binary nanoparticle superlattices is also achieved by co-crystallization of nanoparticles with nanoparticle-loaded protein containers.[190] Gold nanoparticles stabilized with an amphiphilic ligand can co-crystallize with ferritin to obtain highly ordered structures. The incorporation of iron oxide nanoparticles inside the ferritin cavity did not alter the structure of the nanoparticle-protein container assembly resulting in a binary superlattice. By variation of the ionic strength different assembly behavior was observed. However, only small domain sizes were obtained because the AuNPs function as building block and their polydispersity hampered the formation of larger assemblies. Moreover, a modularity in the formation of alternative structures was achieved due to alteration of the ionic strength or by exchanging the ferritin protein container into cowpea chlorotic mottle virus, but the control of the composition of the nanoparticle superlattice is limited. Using AuNPs as building block, only binary superlattices with AuNPs as one of the components are possible.

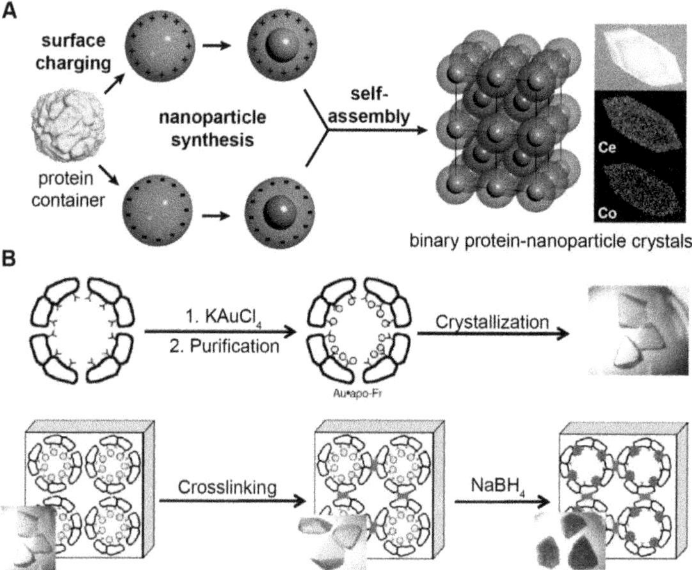

Figure 2.10: Construction of biohybrid materials. (A) Oppositely charged ferritin protein containers with different nanoparticle cargo were assembled by protein crystallization to obtain binary protein-nanoparticle crystals. Figure adopted from reference 40 with permission of the American Chemical society, copyright 2016. (B) Ferritin with gold ions was crystallized and gold nanocluster were synthesized inside the protein crystals by reduction with $NaBH_4$. Figure adapted from reference 191.

The second strategy, namely to first assemble the protein matrix and then load the matrix with inorganic cargo, exploits the high solvent content (25% - 90%) of protein crystals. This porosity enables the diffusion of several compounds for example frequently used in protein crystallography for derivatization techniques.[192] An example for the second strategy was presented by Ueno *et al.*, who synthesized gold clusters inside the protein container cavity of ferritin protein crystals (Figure 2.10B).[191] First, the inner cavity of ferritin was genetically modified to incorporate gold ions. Subsequently, the ferritin mutant with incorporated gold ions was crystallized to obtain highly ordered protein crystals. The stability of the protein crystals was enhanced by cross-linking and the gold ions were reduced to gold clusters by diffusion of $NaBH_4$ through the protein crystal. Successful synthesis of gold clusters was observed by coloration of the crystal from colorless to dark reddish.

In conclusion, the protein container ferritin is an ideal building block for the construction of highly ordered nanoparticle superlattices. Several alternative structures are possible by metal-assisted or electrostatic assembly of the protein container. Moreover, the inner cavity of the ferritin container enables the incorporation of different nanoparticles before or after the assembly of the protein containers into the protein matrix.

3 Basis for the present work

The basis for this work was laid in previous work in our research group. In this chapter, first, previous results will be summarized and then the general strategy for this work will be introduced. The assembly of nanoparticles into a highly ordered material will be realized using an innovative design approach with two oppositely charged ferritin protein containers as building blocks. The strategy can be subdivided into three steps (Figure 3.1).

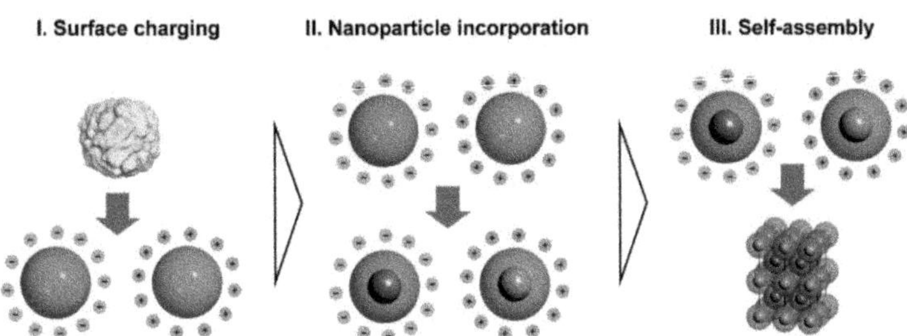

Figure 3.1: Strategy for the formation of nanoparticle superlattices with protein containers as building blocks. (I) Modification of the outer surface charge to obtain two different charged protein container variants. (II) Incorporation of nanoparticles inside the protein container cavity. (III) Self-assembly of the building blocks by protein crystallography.

First, the surface charge of the ferritin protein container is modified to obtain a positively and negatively charged variant. In a second step, nanoparticles are incorporated inside the protein container cavity. Finally, the nanoparticle-loaded protein containers are assembled by protein crystallization in a defined arrangement to obtain binary nanoparticle superlattices, similar to inorganic salts. Importantly, using the same building blocks, the composition of the crystals (unitary or binary) and the lattice parameters can be tuned simply by adjusting the crystallization conditions.

3.1 Supercharging of the protein container ferritin

As building blocks for the nanomaterial two oppositely charged ferritin variants were generated based on the supercharging method. The term supercharging was introduced by Lawrence *et al.* in 2007, where the solubility and aggregation resistance of the green fluorescent protein (GFP) was enhanced by changing the net charge significantly without abolishing their folding or function.[193] The net charge was altered by mutation of highly solvent-exposed amino acids on the surface, i.e. amino acids with the lowest number of neighboring atoms. This method was further improved in 2012 by Miklos *et al.*[194] using the Rosetta design software.[195,196] The program comprises two primary components: an energy function for evaluating the favorable energy of a sequence and an optimization procedure. The energy function considers different parameters such as van der Waals interactions, electrostatic interactions, solvation, hydrogen bonding and a reference term. The reference term includes different values for each amino acid type. In this way, only energetically beneficial amino acids were chosen in contrast to the method of Lawrence *et al.*, where solvent-exposed amino acids were mutated. The surface charging of the ferritin protein container involved two steps. The first step was the computational design of the protein container *via* the Rosetta molecular modelling software. Residues for mutation were chosen based on the energy function of the Rosetta software and manual validation. Negatively and positively charged ferritin containers were designed by substituting surface residues with either aspartic/glutamic acid or with lysine/arginine (Figure 3.2, Appendix Table 8.1).

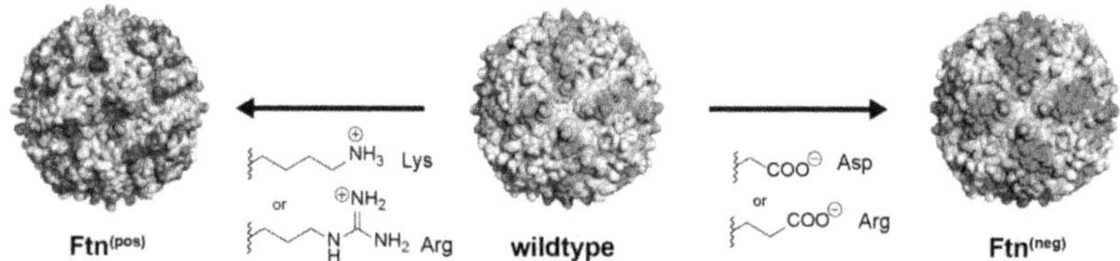

Figure 3.2: Electrostatic surface potential of the ferritin variants. The surface charge was altered by mutation of amino acids on the outer surface (red: -5 kT/e, blue: +5 kT/e).[40]

Mutations considered high risk, e.g. close to the oligomeric interfaces, were excluded. In second step, the charged variants were produced recombinantly in *E. coli.* Protein purification was carried out through ion exchange and size exclusion using fast protein liquid chromatography (FPLC). Two oppositely charged protein ferritin variants, Ftn(pos)[148] and Ftn(neg),[40] were obtained in high purity.

3.2 Nanoparticle synthesis

The incorporation of metal oxide nanoparticles was performed by synthesis of the nanoparticles in the protein container cavity (Figure 3.3). As metal oxide nanoparticles, FeOx,[197] Co_3O_4[198,199] and CeO_2[200] were synthesized similar to the protocol of Uchida *et al.*[32] with minor changes. In detail, the precursor agent and the oxidation reagent diffused through the protein container pores inside the cavity, where the nanoparticle is formed. Thus, the cavity functions as reaction vessel and confines the size of the nanoparticle.

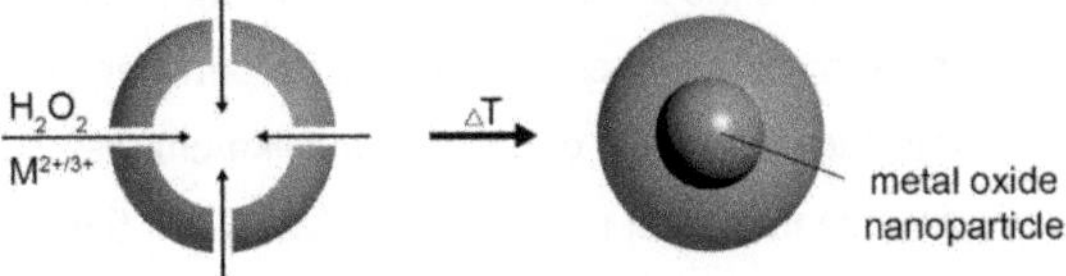

Figure 3.3: Metal oxide nanoparticle synthesis in the protein container cavity. Hydrogen peroxide and metal precursor diffuse through the pores and form the metal oxide nanoparticle.

3.3 Self-assembly

The last step is the self-assembly of the building blocks to obtain the highly ordered material. First, the crystallization of empty protein containers was investigated to study the formation of the protein matrix. In total, four different crystal structures were determined by X-ray crystallography (Figure 3.4).[40,189,201]

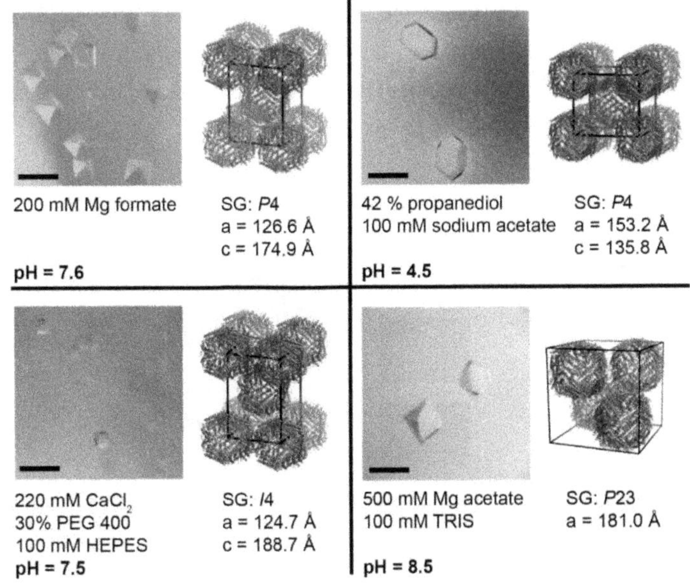

Figure 3.4: Overview of different crystal structures. The crystallization of the oppositely charged protein containers yielded to two binary (top) and two unitary (bottom) structures.[40,189,201]

Interestingly, the composition and the lattice parameters of the protein crystals can be tuned by changing the crystallization condition. In general, the formation of a binary protein crystal composed of both oppositely charged protein containers is favored. A tetragonal lattice is formed whose lattice parameters can be adjusted by changing the pH value from 7.6 to 4.5. Due to the different pH value, the surface charge of the protein containers changed so that different crystal interfaces between the building blocks emerged. At pH 7.6, interfaces between oppositely and like-charged protein containers are formed resulting in a coordination number of twelve (PDB: 5JKL). In contrast, the repulsion of like-charged protein containers at pH 4.5 led to a coordination number of eight and only interfaces between oppositely charged protein containers (PDB: 6H6T). A unitary structure can be formed by complexation of $Ftn^{(neg)}$ with a higher magnesium ion concentration (PDB: 5JKK) or precipitation of $Ftn^{(neg)}$ with calcium ions so that only crystal contacts between $Ftn^{(pos)}$ can emerge (PDB: 6H6U).

Finally, metal oxide nanoparticle-loaded protein containers were utilized to obtain nanoparticle superlattices (Figure 3.5). The composition of these crystals was investigated by scanning electron microscopy (SEM) images. The protein crystals contained both metal oxide nanoparticles. Moreover, small angle X-ray scattering (SAXS) of an assembly of the nanoparticle-loaded crystals revealed that the nanoparticle order is the same as for the empty protein crystal. Thus, the structure of the nanoparticle superlattice is predetermined by the protein matrix.

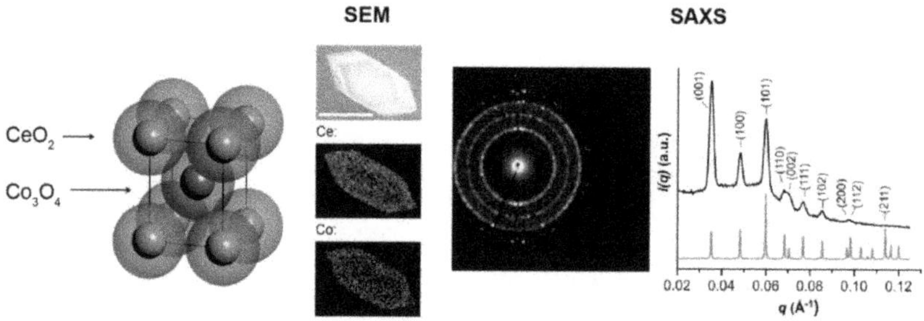

Figure 3.5: Characterization of the nanoparticle superlattice. Binary protein crystals loaded with metal oxide nanoparticles were characterized by scanning electron microscopy (SEM) and small angle X-ray scattering (SAXS). Scale bar is 25 µm. Figure adapted from reference 40 with permission of the American Chemical society, copyright 2016.

4 Concept of the work

The focus in this work will be on the encapsulation of plasmonic nanoparticles and potential applications of the synthesized nanomaterials (Figure 4.1).

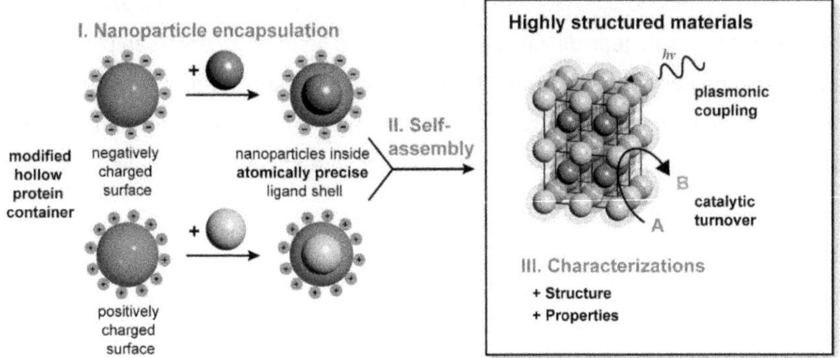

Figure 4.1: General strategy for the construction of nanomaterials composed of plasmonic nanoparticles and protein containers used in this work.

The state of the art is the synthesis of metal oxide nanoparticles in the ferritin protein container cavity. These nanoparticles have potential in catalysis and as components for magnetic materials.[13-15] The synthesis of other nanoparticles such as ZnSe,[35] PbS,[205] calcium carbonate,[206] silver[31] or gold[207] with the nanoreactor approach is possible but individual optimization of each synthesis is required. Furthermore, the fabrication of nanoparticles is limited to syntheses in aqueous solution due to the stability of the protein container. In this work, an alternative strategy for the incorporation of nanoparticles was investigated. An interesting feature of protein containers is the ability to disassemble into their protein subunits at certain conditions and reassemble to reform the protein container after adjusting the original condition.[175,179,208] This feature can be utilized to encapsulate pre-synthesized nanoparticles. The encapsulation of AuNPs in protein containers with different sizes, such as encapsulin,[209] CCMV[210] or MS2 viral capsid[211] was already investigated. In this work, sub-5 nm AuNPs and AgNPs, which have unique plasmonic[212,213] and catalytic[62,70,214] properties, will be synthesized and encapsulated into the modified ferritin protein containers Ftn$^{(pos)}$ and Ftn$^{(neg)}$. In detail, plasmonic nanoparticles with core sizes smaller than 5 nm will be synthesized in organic solvents. Subsequent ligand exchange reactions will afford water-soluble nanoparticles that possess a positively charged outer surface. The encapsulation of the synthesized nanoparticles will be performed with the dis- and reassembly approach. For this purpose, two different conditions will be investigated. In the first condition, the disassembly will be performed in acidic pH (pH = 2.0). After disassembly, the nanoparticles will be added. Reassembly around the particles will be achieved by readjusting the pH to neutral values (pH = 7.4), encapsulating the nanoparticles. The second condition will be a chaotropic condition to unfold and disassemble the protein. High guanidinium concentration will be used for disassembly and reassembly will be performed by lowering the guanidinium concentration (Figure 4.2). The negatively charged inner surface of the ferritin container will favorably interact with the positively charged plasmonic nanoparticles to ensure efficient encapsulation.

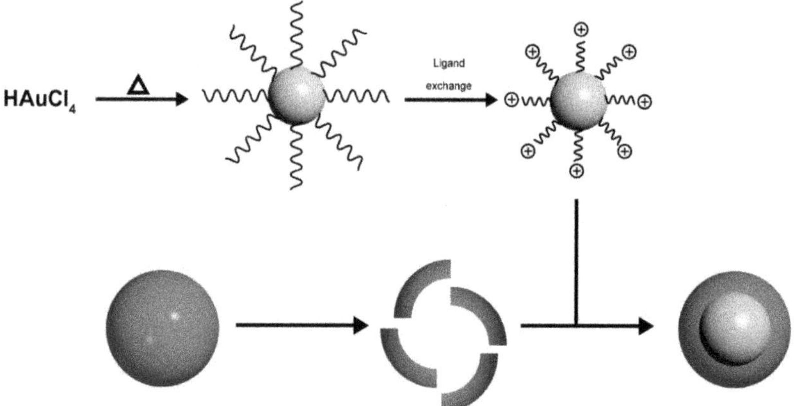

Figure 4.2: Encapsulation of plasmonic nanoparticles. After the synthesis of plasmonic nanoparticles, a ligand exchange reaction will be performed to prepare suitable nanoparticles for the encapsulation.

Self-assembly of the building blocks will be performed by protein crystallization like for the metal oxide loaded protein containers. The structure and the composition of the protein crystals will be studied by optical microscopy, SEM and SAXS experiments. Potential application of the nanomaterial as optical material and catalyst will be investigated. For the study of catalytic properties, plasmonic nanoparticles as well as metal oxide nanoparticles will be utilized.

5 Results

This chapter is divided into three main parts, following closely the strategy outlined for the formation of nanomaterials with protein-nanoparticle composites. As building blocks, positively and negatively surface charged variants of the ferritin protein container are used. The first part addresses the incorporation of the nanoparticles inside the protein container cavity either by encapsulation of plasmonic nanoparticles with the help of the dis- and reassembly approach (chapter 5.1) or the synthesis of metal oxide nanoparticles *in situ* inside the protein container cavity (chapter 5.3). In addition to silver nanoparticles, dye molecules are encapsulated (chapter 5.2). In the second part, the assembly of the protein containers into a highly ordered material is investigated (chapter 5.4). The yield of crystals is increased by optimization of the batch crystallization technique (chapter 5.4.1). Furthermore, the assembly of the newly obtained protein-nanoparticle composites into the highly ordered material is studied (chapter 5.4.2). Finally, the optical (chapter 5.5.1) and the catalytic (chapter 5.5.2) properties of the obtained nanomaterials are investigated.

5.1 Encapsulation of plasmonic nanoparticles

The encapsulation of plasmonic nanoparticles inside the cavity of the negatively surface charged ferritin protein container $Ftn^{(neg)}$ as well as the positively surface charged protein container $Ftn^{(pos)}$ involves three major steps:

1. Synthesis of nanoparticles with a narrow size distribution and subsequent ligand exchange reaction to obtain positively charged nanoparticles.
2. Investigation of the dis- and reassembly behavior of the protein containers.
3. Optimization of the encapsulation procedure and upscaling of the sample amount for crystallization experiments.

In this chapter, the synthesis of AuNPs and AgNPs will be discussed as plasmonic nanoparticles for the encapsulation into the protein container cavity. First, two different AuNP syntheses were optimized and several ligand exchange reactions were performed to obtain a positively charged AuNP. Furthermore, the ligand shell of the most suitable AuNP was studied using several methods. For AgNPs, different syntheses were optimized and stability tests were performed. The dis- and reassembly procedure was investigated for both protein containers. A chaotropic and a pH-dependent condition were inspected. Finally, the encapsulation of AuNPs was performed and the amount of sample was upscaled for crystallization experiments. The results from experiments with AuNPs were later utilized for the encapsulation of AgNP.

5.1.1 Gold nanoparticle cargo

An important prerequisite for the formation of highly ordered materials are monodisperse building blocks. The ferritin container represents an ideal building block due to its atomically precise protein shell. Any inhomogeneity of the nanoparticle can be overwritten by encapsulation of the nanoparticle inside the protein container cavity. However, nanoparticles larger than the protein cavity should be avoided to prevent a deformation of the protein container and thus affect the monodisperse size distribution of the building blocks. In total, three main requirements exist for the AuNP: (I) The inner cavity of ferritin has an inner diameter of 6 nm (Figure 5.1A). In addition to the size limitations of the cavity, the nanoparticle must be stabilized by ligands, which will interact with the inner surface of the protein container. For further applications, the nanoparticles should show a LSPR. The position and intensity of the LSPR depends on the size of the AuNP and no significant LSPR is visible for nanoparticles smaller than 2 nm.[215,216] Regarding all these limitations, the optimal nanoparticle size for encapsulation into ferritin containers should be between 3 nm and 4 nm.

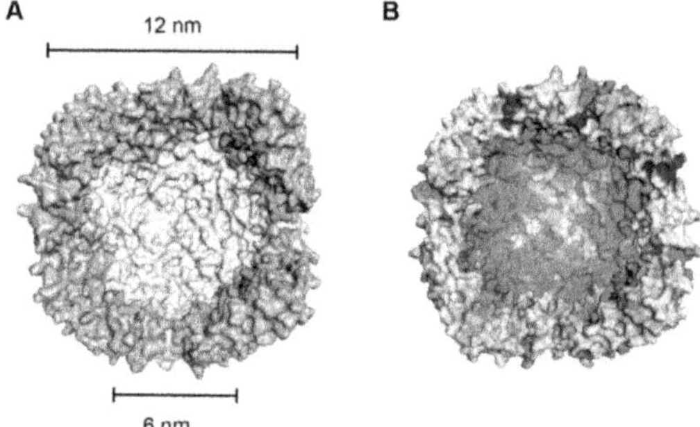

Figure 5.1: Inner cavity of the ferritin protein container. (A) Cut through the ferritin protein container. The inner cavity has a diameter of 6 nm and the outer diameter of 12 nm. (B) Electrostatic potential of the cavity (red: -5 kT/e, blue: 5 kT/e). The inner cavity is negatively charged.

(II) The inner cavity of ferritin is negatively charged (Figure 5.1B). Therefore, the AuNP should be positively charged to enable an interaction with the inner cavity for a high loading efficiency. To fit as much inorganic cargo into the container as possible, short ligands are favored. (III) The AuNP must be soluble in water because the dis- and reassembly is performed in aqueous solutions. Moreover, a certain nanoparticle stability is mandatory to prevent nanoparticle precipitation during the encapsulation procedure.

5.1.1.1 Gold nanoparticle synthesis

In general, AuNPs are synthesized by reduction of chloroauric acid with a reducing agent in the presence of stabilizing ligands. The most common synthesis for water-soluble AuNPs is the Turkevich synthesis.[61] With this method, nanoparticles are obtained in a size range of 9 nm to 120 nm. There are several improvements of this method to gain better control of the particle size,[45,217–220] but it is still difficult to synthesize particles with a size smaller than 5 nm. Brust *et al.* introduced a synthesis of AuNPs by a two-phase system with thiol molecules as stabilizer and sodium borohydride as reducing agent.[75,221] Here, nanoparticles are accessible in a size range from 1.5 nm to 5 nm by changing the ratio of ligand to precursor. Unfortunately, the synthesis was not suitable for this work because the AuNPs are stabilized with thiol ligands. Due to the strong affinity of sulfur to gold, a subsequent ligand exchange is hampered.[70] For that reason, an improvement of the synthesis introduced by Leff *et al.* (samples are abbreviated AuNP-Leff)[76] was investigated. The reaction protocol was slightly modified using dodecylamine as stabilizing agent, which simplifies the ligand exchange afterwards. In this synthesis, the nanoparticle size is mainly influenced by the ligand and not by the reduction speed.[69,222] Therefore, the ligand to gold ratio (L8-L11, Table 5.1) was varied to obtain an optimal size for the AuNPs.

Table 5.1: Summary of data for the AuNP-Leff including UV-Vis measurements and size determination by DLS and TEM.

Sample	Au:ligand	λ_{max} [nm]	d_{DLS} [nm]	PDI	d_{TEM} [nm]
L8	1:8	518	10.38	0.451	4.2 ± 1.3
L9	1:9	517	11.32	0.517	4.5 ± 1.4
L10	1:10	516	11.30	0.219	4.3 ± 0.8
L11	1:11	514	8.21	0.718	3.9 ± 0.9

UV-Vis spectra (Figure 5.2A) of the four approaches showed distinct plasmon absorbance maxima between 514 nm and 518 nm. The position of the maxima was shifted to smaller wavelength by increasing the number of ligands, which indicates a smaller particle size. The volume-weighted dynamic light scattering (DLS) measurements exhibited single peaks for all samples with almost no aggregation (Figure 5.2B) but relatively high polydispersity index (PDI) for all samples except of L10 (Table 5.1). The transmission electron microscopy (TEM) image and the corresponding histogram of L11 are shown in Figure 5.2C and Figure 5.2D. Spherical AuNP with a size of 3.9 ± 0.9 nm were obtained. The majority of the AuNPs are suitable to fit into the cavity of the protein containers, with just 11% of the nanoparticles being larger than 5 nm.

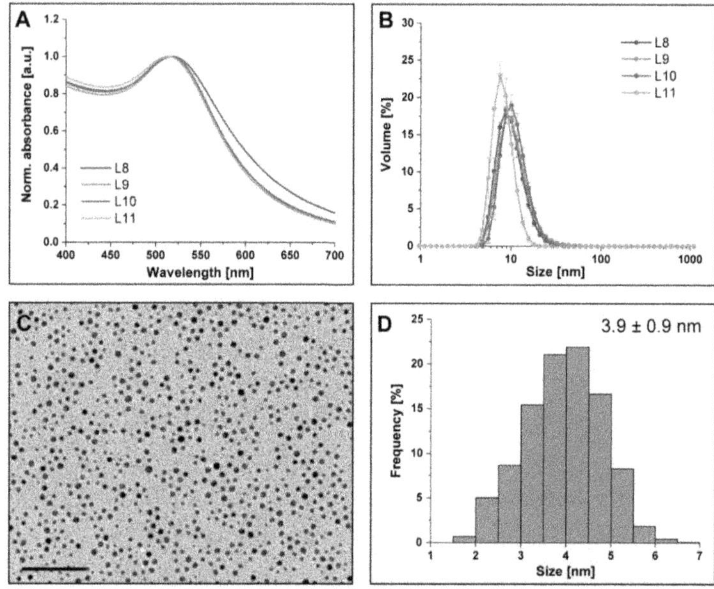

Figure 5.2: Characterization of AuNP-Leff. (A) Normalized absorbance spectra of L8 (black), L9 (red), L10 (blue) and L11 (green). (B) Volume-weighted distribution of the hydrodynamic diameter of L8 (black), L9 (red), L10 (blue) and L11 (green). (C) TEM image and (D) corresponding histogram of L11 with a mean core diameter of d = 3.9 ± 0.9 nm. Scale bar is 50 nm.

To obtain AuNPs with a better size distribution for the encapsulation, a second synthesis introduced by Peng *et al.*[223] (AuNP-Peng) was investigated. Here, AuNPs were synthesized by reduction of chloroauric acid dissolved in 1,2,3,4-tetrahydronaphtalene with *t*-butylamine-borane complex and oleylamine as stabilizing agent. The nanoparticle size can be easily tuned between 2.4 nm and 9.5 nm by decreasing the temperature from 40°C to 2°C. Three different temperatures (25°C, 30°C and 35°C, Table 5.2) were studied.

Table 5.2: Summary of data for the AuNP-Peng including UV-Vis measurements and size determination by DLS and TEM.

Sample	Temperature	λ_{max} [nm]	d_{DLS} [nm]	PDI	d_{TEM} [nm]
P25	25°C	510	4.54	0.171	3.3 ± 0.5
P30	30°C	514	7.11	0.780	3.4 ± 0.4
P35	35°C	518	7.16	0.182	4.7 ± 1.4

The three approaches yield particles that exhibit plasmon absorbance maxima between 510 and 518 nm (Figure 5.3A). The volume-weighted DLS measurements show single peaks for all approaches (Figure 5.3B) with almost no aggregation. Moreover, the PDI has a small value (Table 5.2), indicating a narrow size distribution. TEM image (Figure 5.3C) and the corresponding histogram (Figure 5.3D) of P30 show spherical AuNPs with a mean core diameter of 3.4 ± 0.4 nm.

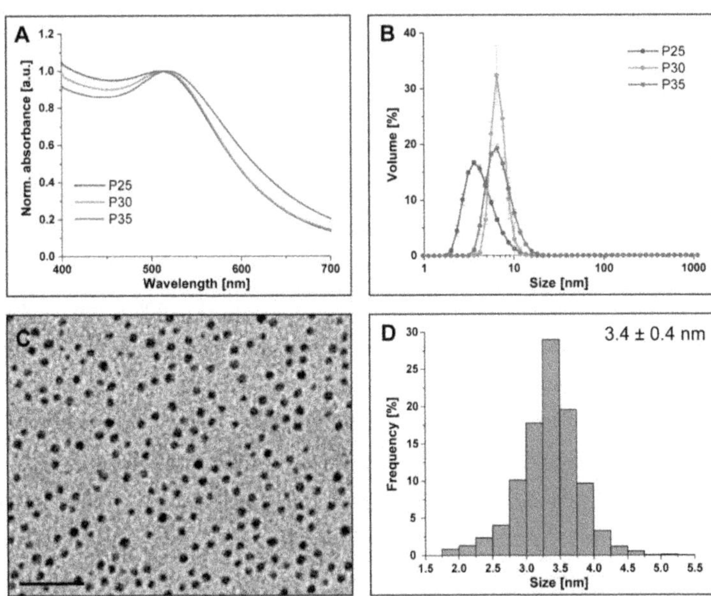

Figure 5.3: Characterization of AuNP-Peng. (A) Normalized absorbance spectra of P25 (black), P30 (red) and P35 (blue). (B) Volume-weighted distribution of the hydrodynamic diameter of P25 (black), P30 (red) and P35 (blue). (C) TEM image and (D) corresponding histogram of P30 with a mean core diameter of d = 3.4 ± 0.4 nm. Scale bar is 25 nm.

Peng *et al.* described a decrease of the AuNP size by increasing the temperature. An increased temperature accelerates the formation of nuclei from the precursor, i.e. more nuclei start growing at the same time. This results in a lower gold atom concentration, when the nuclei start growing and consequently in a smaller particle size. In this work, the opposite effect was observed. At the highest temperature, the largest AuNPs were obtained and the influence of the temperature on the nanoparticle size seems to be not as significant because the samples P25 and P30 showed a similar AuNP size. This behavior could be explained by particle ripening (Ostwald ripening).[64,224] During the Ostwald ripening, smaller particles are dissolved in favor of larger particles, which results in a further growth of the particles. An increased temperature accelerates this process and it seems as this process is faster that the formation of nuclei. In comparison to the AuNP-Leff all AuNPs are smaller than 5 nm (Table 5.2) and should fit into the protein container cavity. Thus, for encapsulation experiments AuNP-Peng was used.

To sum up, the synthesis of AuNPs was successfully performed with two different synthetic routes. Gold nanoparticles (AuNP-Peng) were synthesized with a size between 3 nm and 4 nm. This size fits perfectly for the encapsulation into the protein container cavity. These particles will be used in further experiments within this work.

5.1.1.2 Functionalization and characterization

After the successful synthesis of AuNPs, the next step is the functionalization of the nanoparticle surface. The ligand shell of the AuNP plays a key role for the encapsulation. On the one hand, the ligand shell must be positively charged to ensure an interaction with the inner cavity of the protein container, which is negatively charged. On the other hand, the ligand shell is responsible for the colloidal stability of the nanoparticle in the dis- and reassembly conditions. It shields the AuNPs from destabilizing molecules, which try to adsorb on the particle surface, and prevents agglomeration of particles. The synthesized AuNP-Peng are stabilized with oleylamine, where the ligand binds via an amine functionality to the gold surface. For the ligand exchange reaction, thiol-based ligands[225] are favored because of the higher affinity of sulfur to gold in comparison to nitrogen to ensure complete ligand exchange. In the end, AuNP must be soluble in water since water is the reaction medium for the encapsulation procedure. Towards this end, only two terminal groups apply all requirements: tertiary[71,77] and quaternary[226] ammonium-terminated thiols. The positive charge of the tertiary ammonium-terminated thiols depends on the pH value. The ligand shell is only positively charged, when the amine group is protonated. In contrast, the quaternary ammonium-terminated thiols are permanent positively charged, independently of the pH value.

Moreover, the pH-independent electrostatic repulsion ensures a stronger stability of the AuNPs. Therefore, ligand exchange reactions were performed with quaternary ammonium-terminated thiols. Due to accessibility, ligands with different alkyl chain length were chosen: (11-mercaptoundecyl)-N,N,N-trimethylammonium ($C11^+$), (5-mercaptpentyl)-N,N,N-trimethylammonium ($C5^+$) and (2-mercaptoethyl)-N,N,N-trimethylammonium ($C2^+$) (Figure 5.4).

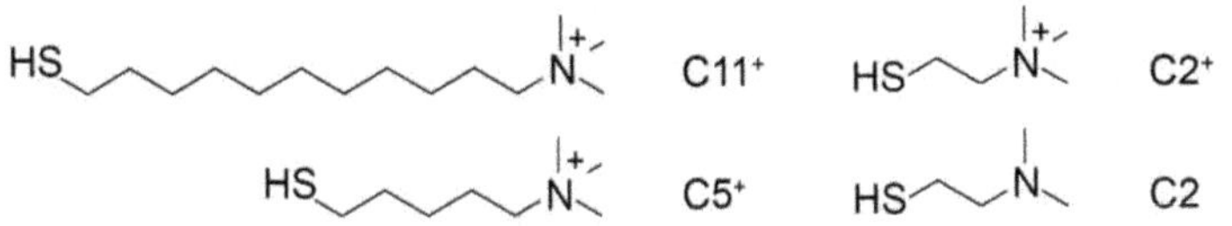

$C11^+$ $C2^+$

$C5^+$ C2

Figure 5.4: Ligands for ligand exchange reaction. Chemical structures of the ligands used for the ligand exchange.

In general, the ligand exchange occurred immediately, and the water phase changed its color from colorless to dark reddish, which indicates a successful ligand exchange. To minimize defects in the ligand shell or incomplete reaction, the ligand exchange reaction was performed overnight and with a large ligand excess. Subsequently, residual ligands were removed by excessive washing with pure water using centrifugal filters. The successful reaction was confirmed by UV-Vis measurements. The ligand exchange reactions with the short ligands $C2^+$ and $C5^+$ were not successful. At the beginning, the colorless water phase turned dark reddish, which confirmed an exchange of ligands, but after a while the nanoparticles precipitated. An increase of ligand excess could not prevent the precipitation of the AuNPs. On the other hand, the AuNPs functionalized with the $C11^+$ ligand ($AuC11^+$) were stable (Figure 5.5).

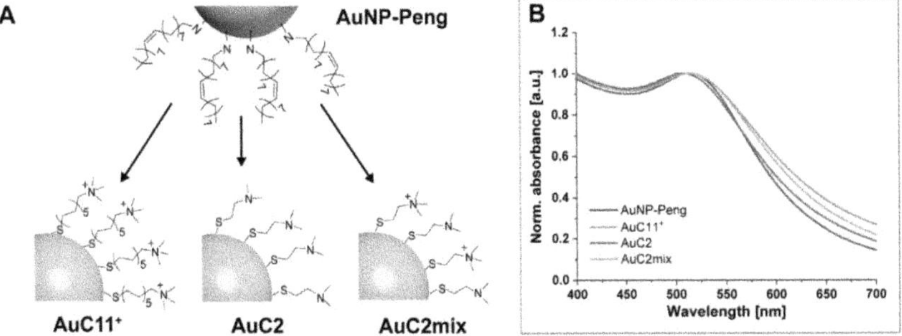

Figure 5.5: Ligand exchange reactions. (A) Ligand exchange reaction of AuNP-Peng to $AuC11^+$, AuC2 and AuC2mix. (B) Normalized UV-Vis spectra of AuNP-Peng (black), $AuC11^+$ (red), AuC2 (blue) and AuC2mix (green).

The only difference between the ligand shells is the alkyl chain length. The length of the ligands is increasing from 534 pm for $C2^+$ (determined by quantum chemical calculation with Gaussian 09[227]) to 1550 pm for $C11^+$.[209] Using the shorter ligands, AuNPs moved from the organic to water phase, which proved the successful attachment of ligands to the AuNP surface. Due to the short alkyl chain, the repulsion between the positively charged quaternary ammonium groups of the ligands possibly leads to an insufficient coverage of the AuNP surface and precipitation of these nanoparticles. In addition to the positive charge, for encapsulation into ferritin containers, the ligand shell thickness should not exceed a certain value because the size of the AuNP is limited by the protein container cavity. As an alternative, the ligand exchange with the short tertiary ammonium-terminated thiol ligand 2-(dimethylamino)ethanethiol (C2) was successfully performed (AuC2, Figure 5.5). During the ligand exchange, the repulsion between the terminal groups is not strong enough to lead to an insufficient coverage of the AuNP surface and subsequent destabilization of the particles. Here, the ligand shell thickness is optimal for the encapsulation, but the charge of the ligand shell could be too weak for an adequate interaction between AuNP and protein container inner surface. Grzybowski *et al.* produced a mixed ligand shell composed of $C11^+$ and the equivalent tertiary amine.[228] The same procedure was transferred to this ligand system with short ligands. The C2 ligand was combined with the $C2^+$ ligand to get a mixed ligand shell (C2mix). In this ligand shell, the C2 ligand has the function of a barrier to prevent strong repulsion between the $C2^+$ ligands, which results in an insufficient coverage of the surface and precipitation of AuNPs. In fact, the mixed ligand shell effectively stabilized the AuNPs (AuC2mix, Figure 5.5).

In addition to the UV-Vis measurement, the AuNPs were further characterized by DLS and zeta potential (ζ-pot.) measurements (Table 5.3). Both measurements are in accordance with the presumed properties. The highest hydrodynamic radius (16.12 nm) was measured for the thickest ligand shell ($AuC11^+$). It has twice the size of the AuNP stabilized with the shorter ligands due to a longer ligand length (534 pm vs. 1550 pm). Moreover, the size determined by DLS differs between AuC2 and AuC2mix, which confirms that the surface of AuC2mix is covered by two different ligands. The difference in size can be explained by the higher charge due to additional $C2^+$. The higher charge results in a larger solvent shell so that a larger size is measured by DLS. The ζ-pot. is increasing in the order $C2 < C2mix < C11^+$, according to the increasing number of quaternary ammonium terminal groups in the AuNP ligand shell.

Table 5.3: Summary of data for the AuNP obtained after the ligand exchange reaction including UV-Vis measurements, size determination by DLS and ζ-pot. measurements.

Ligand	λ_{max} [nm]	d_{DLS} [nm]	PDI	ζ-pot (mV)
C11+	513	16.12	0.477	32.6 ± 3.4
C2	508	6.66	0.139	23.2 ± 3.2
C2mix	514	8.62	0.215	27.0 ± 1.8

By variation of the composition between C2 and $C2^+$, the charge of AuC2mix should be tunable. Therefore, four different compositions were prepared and the ζ-pot. was measured (Table 5.4). As expected, the ζ-pot. increased by increasing the amount of utilized $C2^+$.

Table 5.4: ζ-pot. of AuC2mix with different used compositions for the ligand exchange.

C2:$C2^+$	ζ-pot (mV)
0.25:1	29.8 ± 2.9
1:1	27.0 ± 1.8
2:1	27.0 ± 1.7
4:1	25.9 ± 1.8

The ligand shell composition of AuC2mix can be easily tuned by variation of the ligand concentrations before the ligand exchange reaction. To determine the effective ligand shell composition on the AuNPs, AuC2mix was further characterized by surface-enhanced Raman spectroscopy (SERS),[229,230] nuclear magnetic resonance (NMR)[231] and thermogravimetric analysis (TGA).[72,232] Due to the required large sample quantity for NMR and TGA measurements, first quantification of the ligand shell was performed with SERS because this method is able to analyze small quantities of sample. Raman is a spectroscopic technique used to determine vibrational modes of molecules.[233] Depending on the type of vibration, the wavelength of the scattered light and as a consequence the energy of the vibration differs. This effect can be reported as Raman shift in a spectrum and has characteristic positions for different vibrations. SERS is a special variant of Raman spectroscopy to enhance the intensity of the signals.[234,235] Both ligands have a nearly identical chemical structure. The only difference is an additional CH_3-group and a positive charge at the nitrogen atom for $C2^+$ and a free electron pair at the nitrogen atom for C2. In general, there are no significant differences in the characteristic signals between a tertiary and a quaternary amine in Raman spectroscopy. The electronic environment of the nitrogen atom (positive charge vs. free electron pair) will influence the vibrations in the fingerprint area (below 1500 cm^{-1}), e.g. the position of the binding vibration of aliphatic chains (δ_{C-C}) and the stretching vibration of aliphatic chains (ν_{C-S}). Due to the quantity of signals in this area, the signals overlap and no differentiation between the two ligands is possible. Three areas for the ligands, in which peaks can be distinguish for different ligands, are indicated in the Raman spectra (Figure 5.6A, highlighted in grey). Subsequently, AuC2mix was measured (Figure 5.6B). In comparison to the pure ligands, only few very weak peaks were detected. Nevertheless, it was possible to assign these peaks to the corresponding ligands (a: C2, b: $C2^+$). As a result, both ligands are attached to the AuNP surface, but no prediction of the composition was possible.

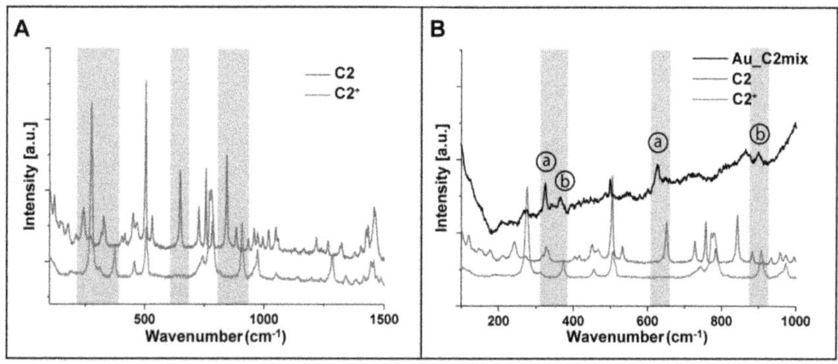

Figure 5.6: SERS measurement of AuC2mix. (A) Raman spectra of the pure ligands C2 (blue) and $C2^+$ (red). The areas are highlighted in grey, in which peaks for the determination of the corresponding ligand can be distinguish. (B) Close-up of Raman spectra of the pure ligands C2 (blue), $C2^+$ (red) and AuC2mix (black). Areas for assignment of peaks to ligands are highlighted in grey. Peaks assigned to C2 are marked with a and to $C2^+$ with b.

The next step was the investigation by ^{1}H-NMR. Typically, NMR is a powerful method to determine the chemical structure of a molecules.[236,237] Although the difference between both ligands is marginal, both ligands should be easy to distinguish. The different electronic composition of the nitrogen atom in both ligands should result in distinct chemical shifts. Figure 5.7 shows the ^{1}H-NMR spectra of the pure ligands. For C2 the ^{1}H-NMR spectrum shows two signals. The triplet at 3.37 ppm belongs to the CH_2-group next to the sulfur. Due to the higher electronegativity of sulfur, the signal has a higher chemical shift compared with the second CH_2-group. The multiplet at 2.93 ppm is an overlap of the triplet of the second CH_2-group and the singlet of the CH_3-groups of the amino group. The integration of the signals confirms the interpretation of the spectrum.

The triplet belongs to two hydrogen atoms and the multiplet to eight hydrogen atoms (two hydrogen atoms from one CH_2-group and six hydrogen atoms from two CH_3-groups). For $C2^+$ five different signals are detected. The two pentets at 2.89 ppm and 3.49 ppm can be assigned to the CH_2-groups. In contrast to C2, a pentet occurs instead of a triplet. A closer look on the pentet reveals that it is a triplet whose central peak split due to rotation of the amino group. The singlet at 3.09 ppm can be related to the three CH_3-groups of the amino group. The other two singlets at 3.13 ppm and 3.20 ppm are impurities. Integration of the signals reveal two hydrogen atoms for each pentet and nine hydrogen atoms for the singlet, which fits with the expected hydrogen quantities for $C2^+$. A comparison of both spectra shows that one signal (triplet) for C2 and two signals (triplet and singlet) for $C2^+$ can be used for the ligand concentration determination. In general, both triplets were chosen for the calculation.

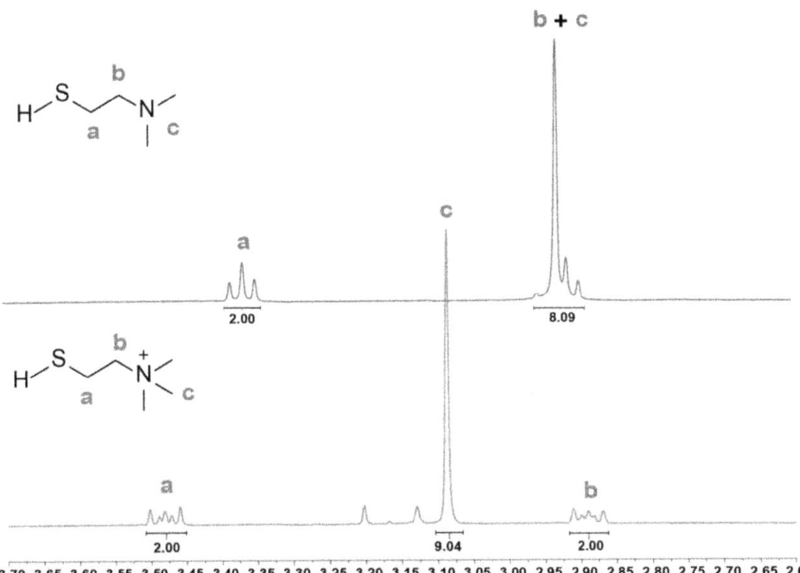

Figure 5.7: ^{1}H-NMR spectra of the pure ligands. The NMR spectra of C2 (blue) and $C2^+$ (red) are stacked. The signals for the chemical groups are marked with letters.

In contrast to the pure ligand, ^{1}H-NMR measurement of AuNPs is more complicated because of signal line broadening. Two main factors are responsible for the line broadening:[238–240] (I) Different gold-thiol binding sites result in a distribution of chemical shifts. This effect becomes weaker with larger distance from the AuNP core. Because of the short ligand length all signals are affected. (II) The slow rotation of the particles due to the large size causes a spin-spin relaxation (T_2) broadening. The dimension of the broadening depends on the AuNP size. The synergy of both factors produces broad indistinguishable signals. However, ligands can be measured with NMR after removal of the inorganic core so that the pure ligand can be detected. In literature, the AuNP core is removed by treatment with sodium cyanide[241] or aqua regia.[242] Both methods were not suitable for this work. By treatment of the AuNP with aqua regia, the removal of the gold core was successful, which was proven by decolorization of the sample. Unfortunately, ^{1}H-NMR spectrum of the sample (Appendix, Figure 8.1) differs from the spectra of the pure ligands. The harsh condition leads presumably to an oxidation of the ligands. However, an alternative strategy was used for the removal of the AuNP core (Figure 5.8).

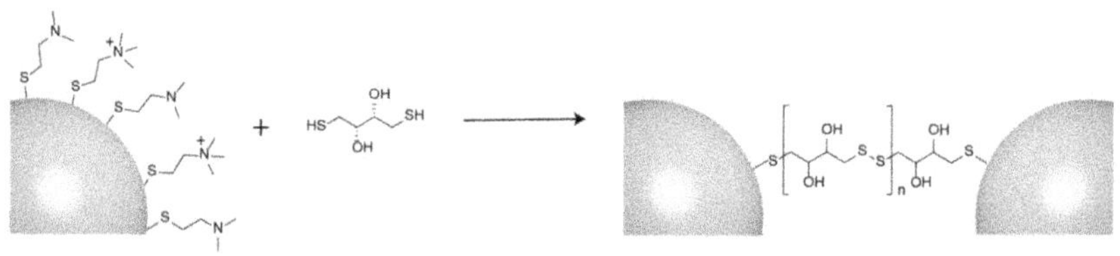

Figure 5.8: Alternative strategy for the determination of the ligand shell composition by ^{1}H-NMR. AuNPs are treated with DTT to exchange the ligands on the surface. Several AuNPs are cross-linked by DTT and precipitate. By centrifugation of the sample, the precipitated AuNP can be easily removed from the solution.

The AuNPs were treated with dithiothreitol (DTT) to perform a ligand exchange reaction. By using this approach, the attached ligands were exchanged with DTT and released into the solution. Moreover, DTT cross-links multiple AuNPs. The AuNPs precipitate and can be easily removed from the solution by centrifugation. Residual DTT does not disturb the ligand concentration determination because their signals in the ^{1}H-NMR spectrum do not overlap with the ligand signals (Appendix, Figure 8.2). A special NMR pulse sequence was used to suppress the water signal and to enhance the ligand signals.[243] Figure 5.9 shows the ^{1}H-NMR spectrum of AuNPs, where C2 to C2$^+$ were applied in a ratio of 1:1 during the ligand exchange reaction. The multiplets at 2.71 ppm and 3.75 ppm belong to residual DDT. In comparison to the ^{1}H-NMR spectra of pure ligands, the signals were shifted to higher values, but the overall positions are as expected. The signals for the CH_2-group near the sulfur atom of both ligands (3.55 ppm for C2$^+$ and 3.37 ppm for C2) were used for the ligand concentration determination. The singlet at 3.16 ppm for C2⁺ was also integrated as control. The integration of the signals shows that the ligands on the AuNP have a 2:1 C2:C2⁺ composition. Although both ligands were used equimolar during the ligand exchange, the twofold number of C2 is attached on the AuNP surface.

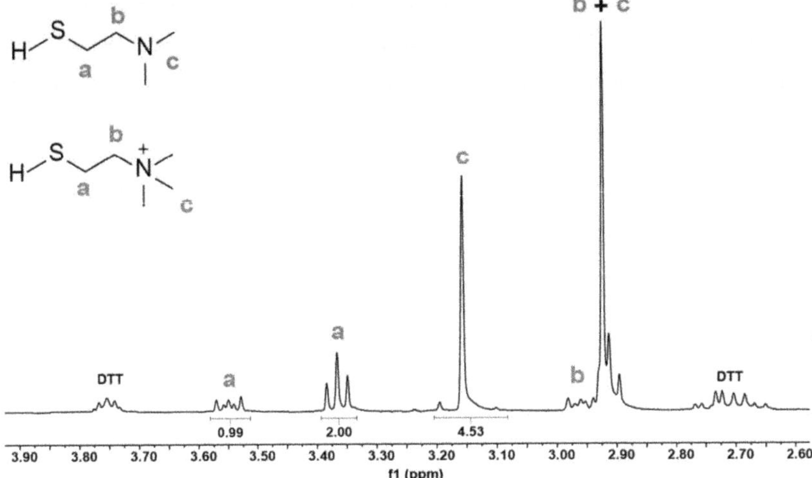

Figure 5.9: ^{1}H-NMR spectrum of an applied ligand ratio of 1:1. The signals for C2 are marked with blue letters and for C2⁺ in red letters. DTT signals are on the left and right edge of the ligand signals.

In total four different approaches (Table 5.5) were investigated. A comparison of the results showed a twice more C2 ratio on the AuNP surface as applied in the ligand exchange.

Table 5.5: Summary of the effective ligand ratio on the AuNP as determined by NMR for four different applied ligand ratios for the ligand exchange reaction.

C2 (expected)	C2⁺ (expected)	C2 (NMR)	C2⁺ (NMR)	NMR/expected
1	4	1	2.71	0.68
1	1	1	0.50	0.50
2	1	2	0.57	0.57
4	1	4	0.56	0.56

The difference between applied and final ligand ratio can be explained by the probability of the ligands to bind to the AuNP surface. The ligand exchange of the pure C2⁺ ligand showed a phase transfer and thus a successful ligand exchange. However, the nanoparticles precipitated due to a presumably incomplete surface coverage of the nanoparticle caused by the repulsion of the positively charged ligand. On the other hand, a successful ligand exchange was performed with the C2 ligand. The C2 is arranged more closely to each other resulting in a higher nanoparticle stability. In the case both ligands are present in the solution, the $C2^+$ ligand can only be arranged in a larger distance to each other, whereas the C2 ligand in much smaller distances close to each other and the $C2^+$ ligand. As a consequence, the number of available binding sites on the nanoparticle surface is higher for the C2 ligand, which leads to a higher surface coverage of the C2 ligand compared to $C2^+$.

The last step for the characterization was the determination of the number of ligands attached to one AuNP. This can be easily investigated by TGA, where the weight loss between 25°C and 400°C corresponds to complete decomposition of the ligands. Instead of the total number of ligands attached to one AuNP, the area which is occupied by one ligand, the ligand footprint (FP), is a better benchmark due to the independency of AuNP size.[232] In literature, the FP was already determined for several similar ligands e.g. aminooctanethiol[244] or $C11^+$.[245,246] In this work, a theoretical and an experimental FP was determined. For the theoretical FP three assumptions were made: (I) The AuNP has a perfectly spherical form. (II) The area covered by the ligand can be described as a circle with the diameter of the largest distance between two hydrogen atoms of the amino group (Figure 5.10A and B). (III) The circles are ordered as close as possible on the AuNP surface (Figure 5.10C). A close packing of equal spheres is assumed, with a coverage of 74%. The size of the AuNP was determined by TEM. To determine the longest distance between two hydrogen atoms of the amino group, a structure optimization was performed (see chapter 7.5.14). The distance was 417.1 pm for C2 and 420.6 pm for $C2^+$.

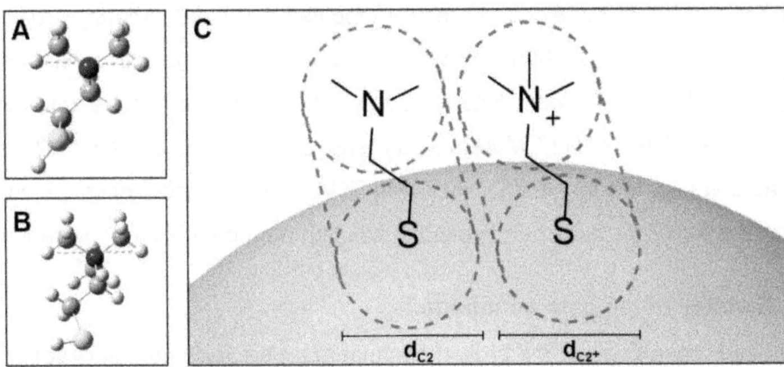

Figure 5.10: Model for ligands on an AuNP surface. (A) Chemical structure of C2. The longest distance between two hydrogen atoms of the amino group is 417.1 pm. (B) Chemical structure of $C2^+$. The longest distance between two hydrogen atoms of the amino group is 420.6 pm. (C) Arrangement of the ligands on the AuNP surface. The coverage is assumed to be 74% of the whole surface area.

Both distances were averaged and an area of 0.14 nm^2 was calculated. Taking the close packing into account, an effective area of 0.19 nm^2 was obtained. For the experimental determination of the FP, TGA measurements of the pure ligands and AuNPs with the ligand shell, AuC2mix, were performed (Figure 5.11).

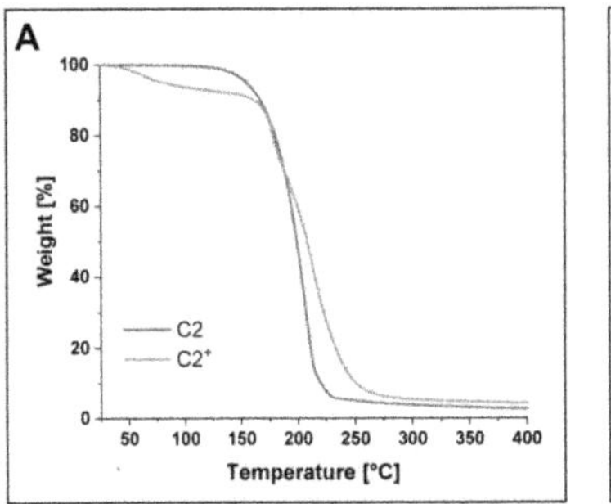

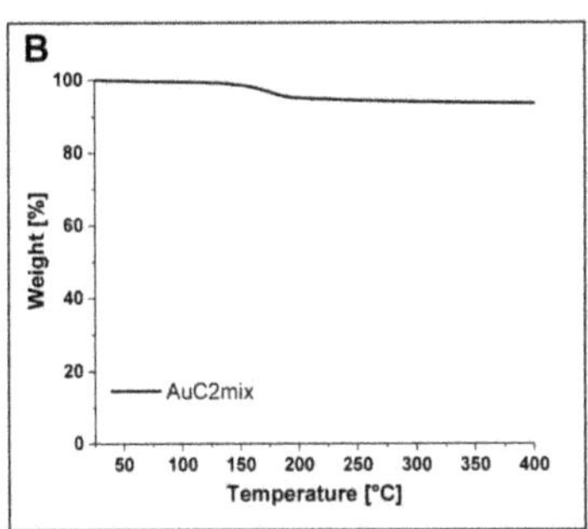

Figure 5.11: TGA measurements of the pure ligand and AuC2mix. TGA measurements of (A) the pure ligands C2 (blue) and $C2^+$ (red) and (B) AuC2mix.

The FP can be readily calculated from the mass loss during the TGA measurement (see chapter 7.5.12). From these values, a footprint of 0.20 nm^2 was calculated. Thus, a similar value as for the theoretical calculation was obtained (0.20 nm^2 vs. 0.19 nm^2). The theoretical calculation does not include interactions between different ligands e.g. repulsion of the terminal amino groups, which results in a higher total number of ligands per AuNP equivalent to a smaller FP. For aminooctanethiol the FP is 0.22 nm^2[244] and for $C11^+$ mixed with PEG ligands 0.27 nm^2,[245] which is similar to the FP of C2mix.

In summary, the AuNP shell can be easily modified by ligand exchange reactions to obtain a positively charged AuNP. In total, three different ligand shells were obtained. The charge of the AuNPs were determined by ζ-pot. measurements and increases in the order C2 < C2mix < C11$^+$. Moreover, the effective composition of the C2mix ligand shell was investigated by SERS, NMR and TGA. A two times higher C2 content is found on the AuNP surface than applied in solution during the ligand exchange.

5.1.1.3 Stability of the gold nanoparticles

The dis- and reassembly of the protein container can be performed in chaotropic condition,[175] or in acidic as well as basic pH conditions.[208] The driving force for the encapsulation is the interaction between the positively charged AuNP and the negatively charged inner surface of the protein container. The strength of this interaction can be influenced by the ionic strength during the encapsulation process. Thus, a certain colloidal stability is required to ensure the stability of AuNPs in solution during the encapsulation process. The stability of AuNPs was investigated by measuring UV-Vis spectra (Figure 5.12A) over a certain time interval. If no change in the spectrum was observed, it indicates stable AuNP. Precipitation of AuNPs is visible by a decrease of the intensity due to lower AuNP concentration. Agglomeration of AuNP leads to a larger nanoparticle size, which results in a red shift in the spectrum. For straight-forward analysis of the data, only the change in plasmon absorbance maximum was monitored. In total, the three ligand shells were studied: C11$^+$, C2 and C2mix. The stability of the AuNP was investigated in 7 M guanidinium (chaotropic dis- and reassembly condition), different pH values (pH dependent dis- and reassembly), different sodium chloride concentrations (ionic strength during the encapsulation) and 1 M DTT (to evaluate ligand binding strength). Figure 5.12B shows the stability of AuNPs in different sodium chloride concentrations and 7 M guanidinium (Gua). All AuNPs are stable in 7 M Gua so that an encapsulation in chaotropic conditions should be feasible. The stability in different NaCl concentrations differs for the different ligand shells. AuC11$^+$ reveals no change in the intensity of the plasmon absorbance maximum until a NaCl concentration of 4 M. The positive charge in combination with the thick shell shields the AuNP very well. Interestingly, both thin ligand shells AuC2 and AuC2mix show different behavior. For both AuNPs a similar decrease of plasmon absorbance maximum is observed for a high NaCl concentration at 1 M. At 0.1 M NaCl concentration AuC2mix shows no significant change but the intensity for AuC2 decrease similar to that for the high NaCl concentration. The permanent positive charge of $C2^+$ in the C2mix ligand shell additionally stabilizes the AuNP. Consequently, low NaCl concentrations do not interfere with the stability of AuC2mix and can be used to adjust the interaction between AuNP and inner protein container surface during the encapsulation. AuC2 should only be used if the encapsulation is possible without the need to vary the ionic strength of the solution.

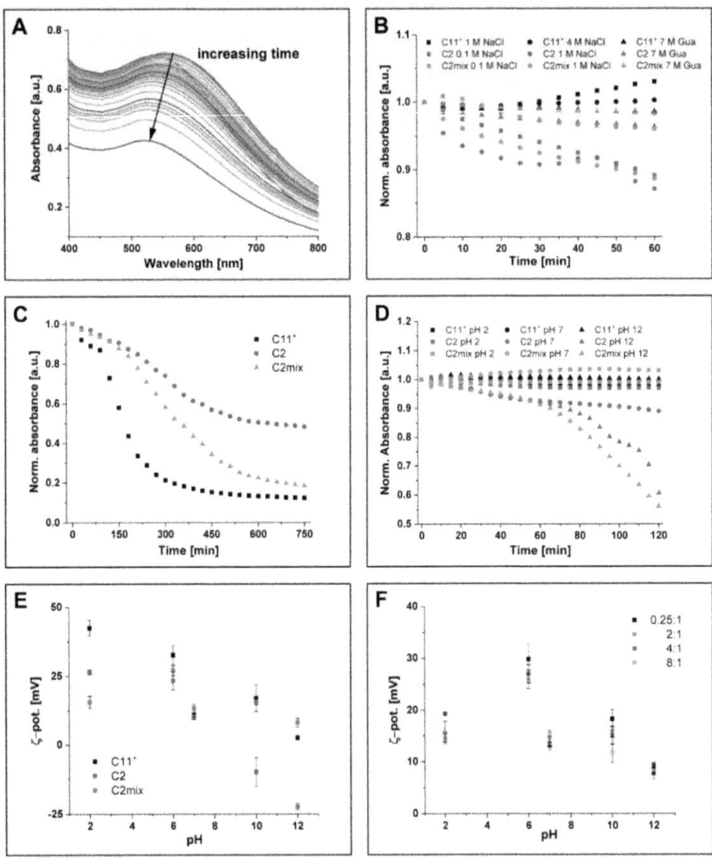

Figure 5.12: Stability test of different functionalized AuNP in different conditions. (A) An example of a stability test measurement. To determine the stability UV-Vis spectra were measured over a certain time interval. Stability of $C11^+$ (black), C2 (blue) and C2mix (red) functionalized AuNP against (B) sodium chloride, (C) ligand exchange with DTT, and (D) pH value. ζ-pot. measurements of (E) different functionalized AuNP and (F) different C2mix ligand shell compositions.

The ligand binding strength (Figure 5.12C) shows a contrary behavior. Here, $AuC11^+$ precipitates very quickly. AuC2mix shows a better stability compared with $AuC11^+$ but still a fast precipitation behavior. The intensity of AuC2 decreased only to half after 750 min, which indicates a high ligand binding strength. It seems that the ligand binding strength of the AuNPs strongly depends on the surface charge. The lowest ligand binding strength is observed for the most positively charged AuNPs. Moreover, despite a similar ligand shell thickness of AuC2 and AuC2mix, a higher stability for AuC2 is monitored confirming the correlation of ligand binding strength and charge. The stability in different pH values (Figure 5.12D) reveals the influence of the permanent positive charge of the ligands for the nanoparticle stability. AuC2 is functionalized with tertiary amines, which are positively charged at high pH values. AuC2mix is partially and $AuC11^+$ permanently positively charged due to functionalization with quaternary amines. All AuNP show no decrease of intensity at pH 2. Moreover, $AuC11^+$ remains unchanged for all pH values. The intensity decreases fast for AuC2mix at pH 12. AuC2 reveals a slow decrease at pH 7 and a fast decrease at pH 12 similar to AuC2mix. All AuNPs can be used for encapsulation in acidic dis- and reassembly conditions. Dis- and reassembly in basic conditions should not be tested because AuC2 and AuC2mix precipitated there very fast. Moreover, AuC2 should be stored in weak acidic conditions because a slow precipitation occurred at neutral pH 7. In addition to the UV-Vis measurements, ζ-pot. measurements were performed to characterize the stability at different pH values in more in detail. The ζ-pot. was measured for different ligand shells (Figure 5.12E) and composition of AuC2mix (Figure 5.12F). Colloidal stability can be classified by ζ-pot. values of $\pm$ 0-10 mV, $\pm$ 10-20 mV, $\pm$ 20-30 mV and $\geq \pm$ 30 mV as highly unstable, relatively stable, moderately stable and highly stable, respectively.[247] This classification considers only electrostatic stabilization and no van der Waals forces,[248] but it is still a good hint for colloidal stability. $AuC11^+$ is highly stable at acidic pH values and relatively stable at pH values lower than 7. AuC2mix is relatively stable for the whole pH range. AuC2 is moderately stable at acidic pH values and highly unstable (negative values) at basic pH values. For different compositions of the ligand shell, no trend or large differences were observed.

Overall, the AuNPs show high stability in Gua and acidic pH values. Basic pH values should be avoided because of the high instability of AuNPs in these conditions. In general, $AuC11^+$ showed the highest stability and AuC2 the lowest.

5.1.2 Silver nanoparticle cargo

The requirements for the AgNPs are the same as for the AuNPs. In summary, the aim is the synthesis of AgNPs with a size between 3 nm and 4 nm, which ligand shell can be readily exchanged to obtain a positively charged nanoparticle. Optimally, the same ligand shell as for AuNPs can be used to have a standardized ligand/nanoparticle system for the encapsulation.

5.1.2.1 Silver nanoparticle synthesis

Although several syntheses for silver nanoparticles with different shapes and size ranges have been published,[63,249,250] it is still difficult to synthesize monodisperse nanoparticle below a size of 5 nm. Generally, similar approaches as for AuNP syntheses are utilized. Leff *et al.* modified their AuNP synthesis for silver to obtain nanoparticles in the size range between 1.8 nm and 3.5 nm.[251] Unfortunately, long alkyl thiols were used as capping ligand, which are difficult to exchange with other ligands. The promising methods are based on thermal decomposition of synthesized silver precursors[252,253] or directly silver nitrate.[254–256] In this work, one synthesis route of both methods was investigated. First, the synthesis of AgNP by Yamamoto *et al.*[252] (AgNP-Yam) was studied. Here, myristic acid was added to an aqueous solution of silver nitrate to synthesize silver myristate. The obtained silver myristate was dissolved in triethylamine and heated up to 80°C for 80 min to obtain the silver nanoparticles. During the thermal deposition, the solution changed its color from grey to brown, indicating the successful synthesis of silver nanoparticles. UV-Vis spectrum (Figure 5.13A) shows a distinct plasmon absorbance maximum at 416 nm. In contrast to AuNPs, a different plasmon resonance frequency occurs because of the different permittivity of the two metals. The size and morphology of AgNPs was further investigated by TEM. Spherical AgNPs with a size of 7.2 nm ± 1.4 nm were obtained (Figure 5.13B). By changing the precursor ligand from myristic acid (C_{14}) to stearic acid (C_{18}), the size of the AgNPs was decreased to 4.8 nm ± 1.4 nm (Figure 5.13C and D). Due to the limited reproducibility of the precursor synthesis, this synthesis was not investigated further.

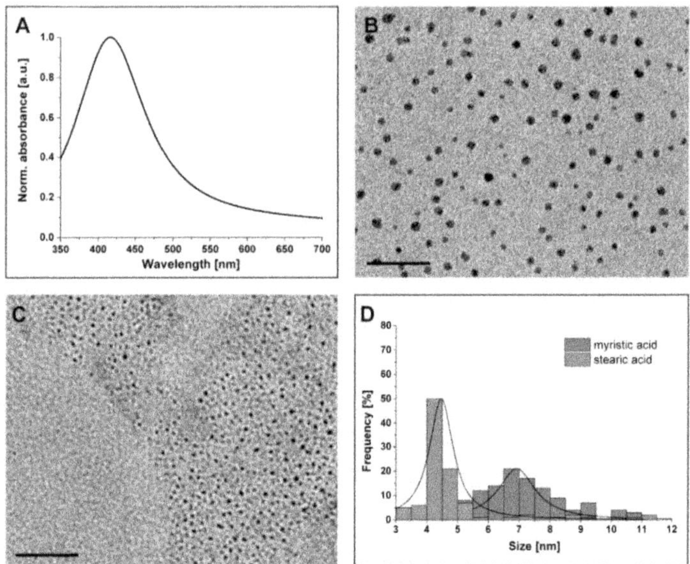

Figure 5.13: Characterization of AgNP-Yam. (A) Normalized absorbance spectrum of AgNP-Yam synthesized with myristic acid as precursor ligand. TEM images of AgNP-Yam with myristic acid (B) or stearic acid (C) as precursor ligand. Scale bars are 50 nm. (D) Corresponding histogram of both syntheses.

In a second step, the synthesis by Hyeon *et al.*[255] (AgNP-Hyeon) was investigated. Here, silver nitrate was thermally decomposed in the presence of oleylamine at 180°C to obtain AgNPs. Interestingly, the size of the nanoparticle can be easily adjusted between 1.7 nm and 3.5 nm by changing the heating rate from 10°C min^{-1} to 1°C min^{-1}. Therefore, four different heating rates (1°C min^{-1}, 5°C min^{-1}, 8°C $min^{-1}$and 10°C min^{-1}) were studied to synthesize the optimal AgNP size. The samples from the different approaches show plasmon absorbance maxima between 423 nm and 436 nm (Figure 5.14A). As expected, the highest heating rate yields the smallest AgNPs so that the plasmon absorbance maximum appears at the smallest wavelength. The TEM image and corresponding size distribution of the 10°C min^{-1} heating rate synthesis showed spherical AgNP with a size of 3.1 nm ± 0.3 nm (Figure 5.14B and C). A comparison of the obtained AgNP sizes revealed the expected behavior. With a lower heating rate, larger AgNPs were obtained. A further increase of the heating rate to 24°C min^{-1} yielded no stable AgNPs. In contrast to the published procedure, the total size of the synthesized AgNPs was larger. Nevertheless, the obtained AgNPs with a 10°C min^{-1} heating rate show an adequate size for encapsulation in the protein container.

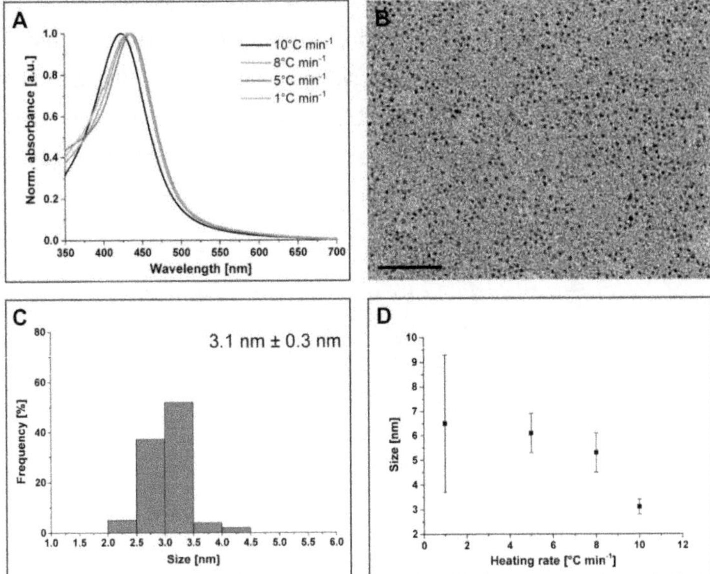

Figure 5.14: Characterization of AgNP-Hyeon. (A) Normalized absorbance spectra of AgNP synthesized with different heating rate. (B) TEM image and (C) corresponding size distribution of AgNP synthesized with a 10°C min^{-1} heating rate with a mean core diameter of d = 3.1 ± 0.3 nm. Scale bar is 50 nm. (D) Dependency of nanoparticle size on heating rate.

As a result, two different AgNPs syntheses by thermal decomposition were investigated. In the first synthesis, silver precursors were synthesized and subsequently thermally decomposed in the presence of a ligand to obtain the AgNP. The obtained nanoparticles had a size of 4.8 nm ± 1.4 nm. Due to the limited reproducibility, the synthesis was not investigated further. The second synthesis used directly silver nitrate in the presence of oleylamine for thermal decomposition. By variation of the heating rate, AgNPs with a size of 3.1 nm ± 0.3 nm were obtained. These AgNPs have an adequate size for encapsulation and will be used for further experiments.

5.1.2.2 Functionalization and stability test

After the synthesis, AgNPs are functionalized with oleylamine to stabilize the nanoparticle. However, a positively charged nanoparticle is required for the encapsulation into the ferritin container. The functionalization of AuNPs by ligand exchange reactions was described in detail in the previous chapter 5.1.1.2. Moreover, AuNP-Peng were also stabilized with oleylamine before the ligand reactions. Therefore, the results for AuNPs can be utilized for AgNPs. In detail, the ligand exchange was performed with $C11^{+}$ as well as C2mix ligands similar to the AuNPs with a minor adjustment. Due to the lower stability of silver compared to gold, the ligand exchange reaction was incubated at 4°C overnight without exposure to light. Again, successful exchange of the ligand shell was monitored by UV-Vis spectroscopy. The UV-Vis spectra of AgNP-Hyeon before and after ligand exchange with C11$^+$ as well as C2mix ligands are shown in Figure 5.15.

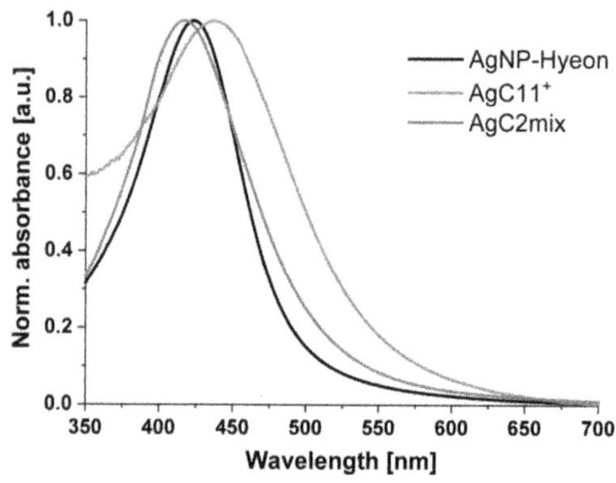

Figure 5.15: Ligand exchange reactions with AgNPs. Normalized UV-Vis spectra of AgNP-Hyeon (black), AgC11$^+$ (red) and AuC2mix (blue).

For all three ligand shells, a distinct plasmon absorbance maximum was visible. Moreover, depending on the ligand shell, the position of the maximum varied. For the shorter C2mix ligand, the plasmon absorbance maximum was shifted to 418 nm and for the larger ligand C11$^+$ to 437 nm.

After the successful ligand exchange reaction, the stability of the AgNPs was investigated. In detail, only the stability of AgC2mix was studied, as the short ligand shell is most suitable for ferritin encapsulation. Similar to the AuNPs, the stability against ionic strength (sodium chloride and guanidinium), pH value and ligand exchange with DTT was monitored by UV-Vis spectroscopy. Moreover, the stability of different C2mix ligand shell compositions was investigated by ζ-pot. measurements. The amount of C2 was varied from 0.5 to 8-fold excess. Figure 5.16A shows the stability of AgC2mix against different sodium chloride concentrations and 7 M Gua.

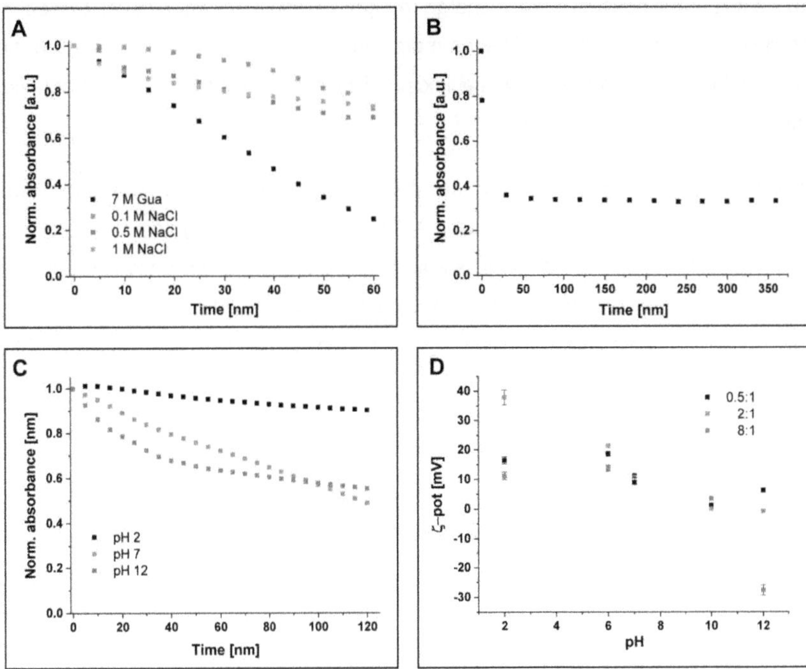

Figure 5.16: Stability tests of AgC2mix in different conditions. Stability of AgC2mix (A) against ionic strength (B) ligand exchange with DTT and (C) different pH values. (D) ζ-pot. measurements of different C2mix ligand shell compositions.

In the 7 M Gua solution, the plasmon absorbance maximum of AgC2mix decreased linear with the measured time. The reduction of the intensity was caused by precipitation of the nanoparticles. The low stability in Gua can lead to difficulties during the encapsulation with this condition. Different sodium chloride concentrations did not have a significant effect on the stability. By treatment with 1 M DTT, AgC2mix precipitated immediately (Figure 5.16B). After 30 minutes, the ligand exchange reaction was finished, and the intensity reached their final value due to cross-linking of the nanoparticles because of the DTT ligand. The measurement in different pH values showed a high stability at acidic pH values (Figure 5.16C). However, at neutral pH as well as basic pH the colloidal stability decreased drastically. The absorbance is linearly dependent on the concentration of the nanoparticles in the solution (Beer Lambert-law). After 120 minutes, half of AgC2mix is precipitated. A similar behavior was observed for different compositions of the C2mix ligand shell. Here, the ζ-pot. was measured (Figure 5.16D). At acidic pH values, a higher colloidal stability was obtained than at basic pH values due to the positive surface charge. A high C2 ratio of the ligand shell decreased the colloidal stability at basic pH even more. However, an increased $C2^+$ ratio did not lead to a higher colloidal stability at acidic pH. The most stable particles were obtained with a balanced C2 to $C2^+$ ratio.

In general, the ligand exchange to get a positively charged AgNP was successful. However, compared with the AuNPs the AgNPs showed a strongly decreased colloidal stability. AgNPs were not colloidally stable in 7 M Gua. Moreover, a high stability was only observed in acidic pH values. Regarding the encapsulation, the low colloidal stability will cause some problems if the AgNPs precipitate during the reassembly of the protein container.

5.1.3 Dis- and reassembly of the protein container

A crucial step for the encapsulation process is the opening of the protein container and closing of it after the synthesized nanoparticles were added. For an optimal encapsulation of nanoparticles, it is important that on the one hand, all protein containers are disassembled, and on the other hand, a complete reassembly of the protein containers takes place. It is essential for the self-assembly of the building blocks that protein containers with the same size are formed again. Although the encapsulation might be successful, an inhomogeneous sample can prevent the crystallization of the sample and as a result the production of the highly ordered material. For the dis- and reassembly two different conditions can be utilized: chaotropic[175], for example using guanidinium hydrochloride (Gua) or pH dependent conditions[208] (Figure 5.17).

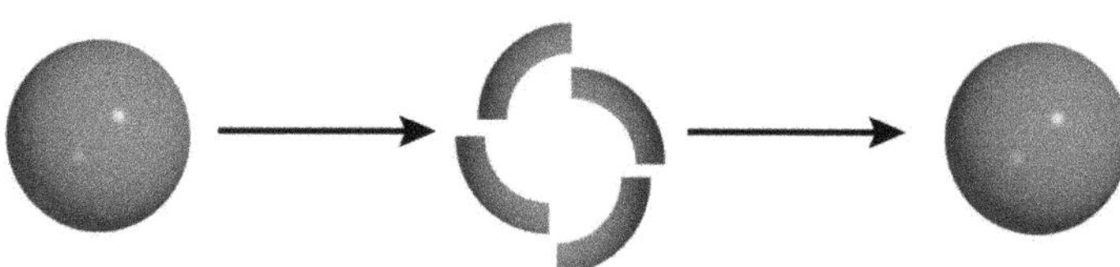

Figure 5.17: Dis- and reassembly of the protein container. Disassembly and reassembly of the protein container in the Gua and pH 2 condition.

The investigation of the nanoparticle stability revealed that Gua and pH 2 are feasible conditions because of the high nanoparticle stability at these conditions. Basic conditions were excluded because of low nanoparticle stability, which will hamper the encapsulation process due to precipitation of nanoparticles. Both conditions, Gua and pH 2, were investigated for both protein container variants. In general, in the first condition, the disassembly was performed at 7 M Gua and reassembly by dilution to 0.5 M Gua. For the second condition, the protein was disassembled in acidic pH (pH = 2) and reassembled by adjusting the pH value back to neutral (pH = 7.4). The dis- and reassembly behavior was monitored by size exclusion chromatography (SEC), circular dichroism (CD) spectroscopy, TEM and DLS. All samples were compared with untreated protein containers. Finally, the samples were crystallized to prove that reassembled samples can organize into the highly ordered material. An unsuccessful crystallization of the treated sample would make an encapsulation of nanoparticles using that condition unnecessary because the crystallization of loaded protein containers is usually more difficult than of empty one.

5.1.3.1 Dis- and reassembly of Ftn$^{(pos)}$

First, the dis- and reassembly in chaotropic condition was investigated. Ftn$^{(pos)}$ was incubated for 1 h, 2 h and 4 h in 7 M Gua at room temperature (Appendix, Figure 8.3) to determine the time when all protein containers are disassembled. After 2 h, no protein containers were visible in the TEM image. It is unclear if residual guanidinium ions can influence the subsequent encapsulation. To minimize this factor, a low concentration was aimed for. Therefore, the disassembly of Ftn$^{(pos)}$ was performed in different Gua conditions to find the minimum Gua concentration for complete disassembly (Appendix, Figure 8.4). A concentration of 7 M Gua is required for complete disassembly of the protein containers (Figure 5.18A). For reassembly, the solution was diluted to 0.5 M Gua and stored overnight at 4°C. A lower temperature was chosen to decelerate the reassembly process and ensure correct formation of the protein container. TEM image of the sample (Figure 5.18B) revealed a successful reassembly of the protein container. The dis- and reassembly process was further characterized by SEC (Figure 5.18C). Here, it was only possible to monitor the sample after reassembly, due to incompatibility of the Gua condition with any column material. Both samples, untreated and reassembled Ftn$^{(pos)}$, showed the same elution volume. The size of the protein container did not change due to the reassembly process. The identical peak width for both samples indicated an unchanged size distribution of the protein containers after reassembly. The amount of reformed protein containers was 19 %.

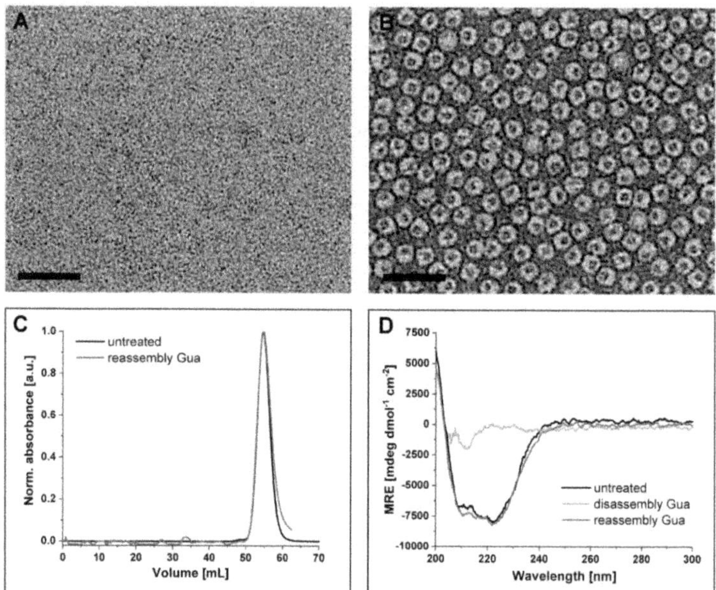

Figure 5.18: Characterization of the dis- and reassembly process of Ftn(pos) in the Gua condition. TEM images of the (A) disassembly in 7 M Gua after 2 h and (B) reassembly after dilution to 0.5 M Gua and incubation overnight at 4°C. Scale bars are 50 nm. (C) Size exclusion chromatogram of the sample after reassembly. The untreated sample (black) has the same elution volume as the reassembled sample (blue). (D) CD spectra of untreated Ftn(pos) (black), after disassembly (green) and after reassembly (blue).

Additionally, the structure of the protein container was investigated by far-UV CD spectroscopy (Figure 5.18D). Far-UV CD spectroscopy can be used to reveal the important characteristics of the secondary structure in proteins. Changes in the secondary structure can be easily detected by a difference in the curve progression in the CD spectra because each secondary structure type shows a characteristic curve. The untreated ferritin protein container showed a typically spectrum with high α-helix content, as seen by the minima at 210 nm and 222 nm. The disassembled sample showed no characteristic curve, which revealed complete denaturation of the protein container. After reassembly, the same characteristic curve is observed as for the untreated protein container. The size of the protein container was determined by TEM and a slightly larger protein container size of 15.0 ± 1.0 nm was measured (Table 5.6). DLS measurement revealed a similar hydrodynamic radius than for the untreated protein container (Table 5.6). Moreover, the low PDI of the reassembled sample indicated a small size distribution of the protein containers.

Table 5.6: Size of the protein container before and after reassembly measured by DLS and TEM.

Sample	d_{DLS} [nm]	PDI	d_{TEM} [nm]
Ftn$^{(pos)}$, untreated	14.57	0.030	14.6 ± 1.7
Ftn$^{(pos)}$, reassembly	14.99	0.078	15.0 ± 1.0

In a second step, the dis- and reassembly in acidic conditions was investigated. For disassembly, Ftn$^{(pos)}$ was incubated for 4 h in phosphate buffer at pH 2. SEC (Appendix, Figure 8.5) of the sample showed complete disassembly of the protein container because no signal is visible at the typically elution volume of the protein container. A reassembly of the protein container was not successful. By adjusting the pH back to neutral values, the protein precipitated completely. A variation of the temperature or adjusting the speed of the pH adjustment did not prevent the precipitation. A higher pH value was not suitable because ferritin shows a stepwise disassembly. A certain pH value (pH = 1.96) is required to ensure complete disassembly.[208] The main difference between Ftn$^{(pos)}$ and wild-type ferritin is the positively charged outer surface. The inner surface of both protein containers is negatively charged. During the reassembly for the wild-type ferritin, the charge of the surfaces plays no key role because both are negatively charged. It seemed that for Ftn$^{(pos)}$, an interaction between monomers occurred: The positively charged outer surface of one monomer might interact with the negatively charged inner surface of another monomer, which leads to an agglomeration of the sample.

The last step was the crystallization of the reassembled sample. As mentioned, it is a crucial step because an unsuccessful formation of the highly ordered material would make the encapsulation pointless. SEC and DLS measurements showed a high purity of the sample so that no further purification of the sample was necessary. The sample was crystallized with untreated Ftn$^{(neg)}$ and binary protein crystals with a size between 80 µm and 120 µm (Figure 5.19) were obtained.

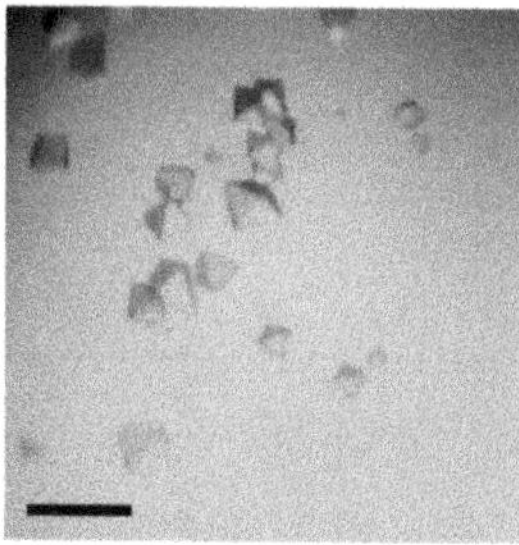

Figure 5.19: Crystallization of reassembled Ftn$^{(pos)}$. Reassembled Ftn$^{(pos)}$ was crystallized with untreated Ftn$^{(neg)}$ to obtain large octahedral crystals. Scale bar is 200 µm.

In conclusion, the dis- and reassembly of Ftn$^{(pos)}$ was successfully performed in the chaotropic guanidinium hydrochloride condition. Well-formed and correctly sized protein containers were obtained suitable for the crystallization into binary crystals with untreated Ftn$^{(neg)}$. However, although complete disassembly was performed in the acidic condition (pH = 2), the reassembly into the protein containers led to precipitation of the protein and thus no reformed protein containers.

5.1.3.2 Dis- and reassembly of Ftn$^{(neg)}$

The investigation of the dis- and reassembly procedure for Ftn$^{(neg)}$ was similar to Ftn$^{(pos)}$. For the Gua condition, the incubation time and Gua concentration were determined to ensure complete disassembly of all protein containers. Here, the protein container must be incubated at 7 M Gua for 4 h (Figure 5.20A). The dilution of the sample to 0.5 M Gua led to complete protein containers (Figure 5.20B). The process was again further characterized by SEC and far-UV CD spectroscopy. The SEC confirmed the formation of protein containers in the same size range (Figure 5.20C). The minor shift in the elution volume compared to the untreated protein container can be neglected. The amount of reformed protein container was 66 %. Far-UV CD spectroscopy showed a complete degeneration of the protein container after disassembly (Figure 5.20D). After reassembly, the characteristic curve is visible but with a smaller α-helix content compared with the untreated protein container.

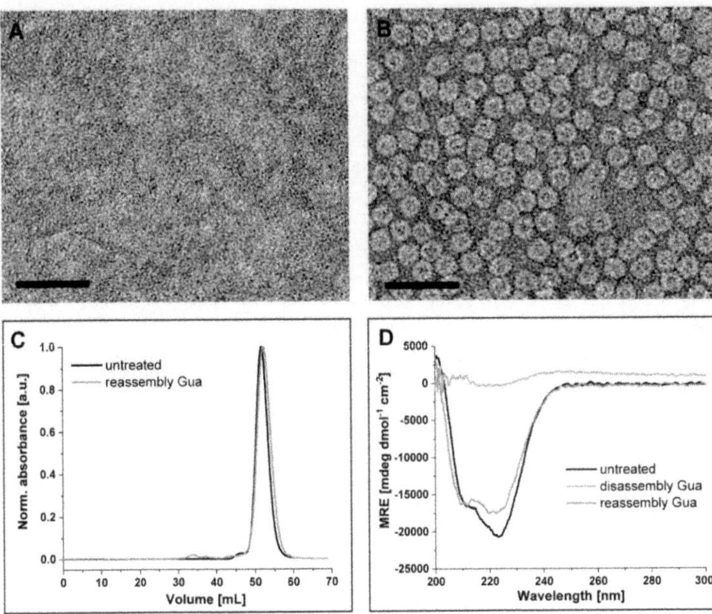

Figure 5.20: Characterization of the dis- and reassembly process of Ftn$^{(neg)}$ in the Gua condition. TEM images of the (A) disassembly in 7 M Gua after 4 h and (B) reassembly after dilution to 0.5 M Gua and incubation overnight at 4°C. Scale bars are 50 nm. (C) SEC of the sample after reassembly. The untreated sample (black) has similar elution volume as the reassembled sample (red). (D) CD spectra of untreated Ftn$^{(neg)}$ (black), after disassembly (green) and after reassembly (red).

Additionally, the dis- and reassembly of Ftn$^{(neg)}$ in the acidic condition was investigated. Compared with the positively charged variant, Ftn$^{(neg)}$ is more similar to the wild-type ferritin. For these two containers, the outer and the inner surface are negatively charged, which should not lead to any agglomeration of the protein container monomers as for Ftn$^{(pos)}$ during the reformation of the protein container. The disassembly of the protein container in phosphate buffer at pH 2 and the reassembly by adjusting the pH back to neutral value were successful. TEM image of the sample at pH 2 show no protein containers (Figure 5.21A). After reassembly, protein containers are formed again (Figure 5.21B).

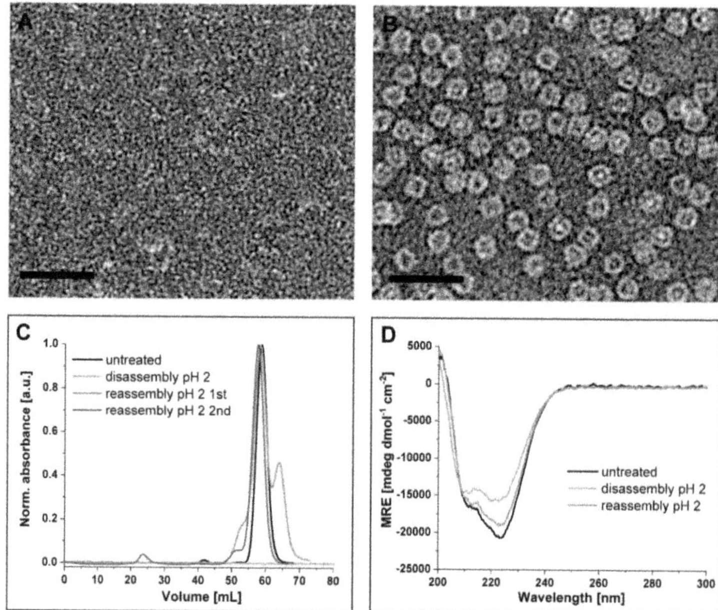

Figure 5.21: Characterization of the dis- and reassembly process of Ftn$^{(neg)}$ in the pH 2 condition. TEM images of the (A) disassembly in pH 2 after 4 h and (B) reassembly after adjusting the pH back to neutral values and incubation overnight at 4°C. Scale bars are 50 nm. (C) SEC of the sample after reassembly. The untreated sample (black) has similar elution volume as the reassembled sample after the first (red) and second (purple) purification. The disassembled sample (green) shows no peak. (D) CD spectra of untreated Ftn$^{(neg)}$ (black), after disassembly (green) and after reassembly (red).

Further characterization was performed with SEC (Figure 5.21C). In comparison to the Gua condition, the disassembly of the sample could be monitored because the condition is suitable for the column. After disassembly of the protein container, no signal is visible in the chromatogram, which indicates the complete disassembly of the protein container. After reassembly at neutral pH, three different peaks appeared in the chromatogram. The main peak shows the same elution volume as for the untreated protein container and contained the reformed protein container. The first peak between 50 mL and 55 mL can be referred to larger species, which are formed by agglomeration of protein containers. The third peak at 65 mL belonged to species, which are smaller than the protein container. Protein container monomers would be expected at much higher elution volumes. Thus, these are probably incompletely formed protein containers. Originally, the pH was adjusted by adding the double amount of phosphate buffer at pH 7.4. The reassembly procedure was optimized, and the pH was adjusted further with NaOH to pH 7.4. This adjustment eliminated the formation of incomplete protein containers (Appendix, Figure 8.6). The amount of reformed protein containers was 60 %. The sample was purified a second time resulting in a chromatogram with mainly one peak. A weak shoulder for agglomerated protein container was still visible. Far-UV CD spectroscopy (Figure 5.21D) showed a different curve for the disassembled sample. The maximum at 222 nm decreased, which indicates a lower α-helix content. After reassembly, the maximum at 222 nm increased again and the curve showed the same characteristic course as for the untreated protein container. The last step was the crystallization of the sample. To evaluate the purity of the sample DLS measurements of the reassembled samples were performed (Table 5.7).

Table 5.7: Size of the protein container before and after reassembly determined by DLS and TEM.

Sample	d_{DLS} [nm]	PDI	d_{TEM} [nm]
$Ftn^{(neg)}$, untreated	13.50	0.038	15.2 ± 1.2
$Ftn^{(neg)}$, reassembly Gua	13.35	0.031	15.3 ± 2.0
$Ftn^{(neg)}$, reassembly pH 2	19.04	0.313	14.1 ± 1.3
2nd SEC	14.73	0.019	

The reassembled sample with the Gua condition has quite the same hydrodynamic radius and PDI as the untreated protein container. Here, the dis- and reassembly has no effect on the sample quality. Moreover, there was no difference for the protein container size determined by TEM. For the reassembled sample with the pH 2 condition, two SEC purifications were needed to obtain a monodisperse sample confirmed by the DLS results. The first SEC resulted in a high hydrodynamic radius of 19.04 nm and a PDI of 0.313. The second SEC reduced the PDI to 0.019, which is more suitable for crystallization. A hydrodynamic radius of 14.73 nm was obtained, which was still higher than for the untreated protein container. Surprisingly, the size determined by TEM was smaller than for the untreated sample. The crystallization was possible for all samples (Figure 5.22). For the reassembled sample with the Gua condition, large octahedral crystals were obtained. As expected, the crystallization of the reassembled sample with pH 2 condition after 1st SEC led to crystals with many defects because of the insufficient purity of the sample. The 2nd SEC improved the purity of the sample which resulted again in octahedral crystals.

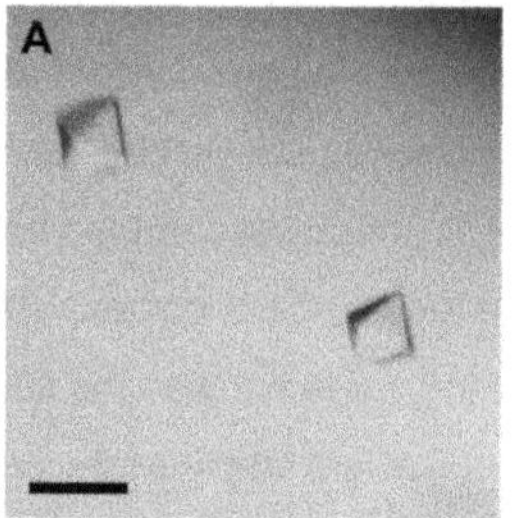

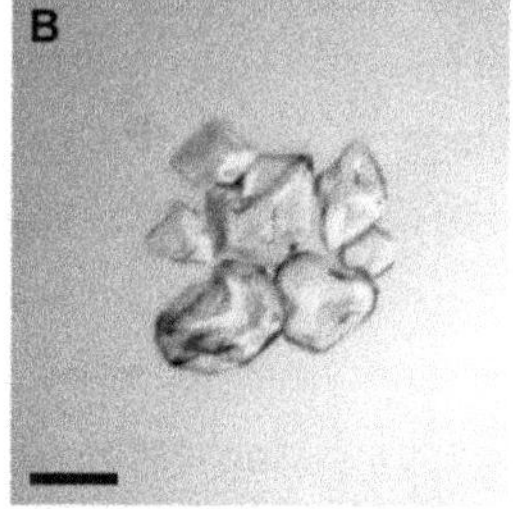

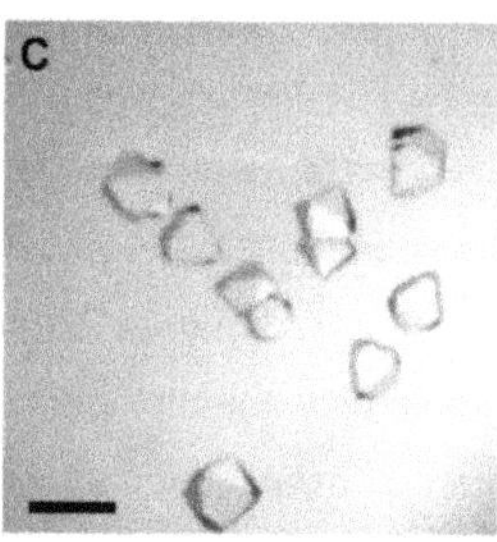

Figure 5.22: Crystallization of reassembled $Ftn^{(neg)}$. Reassembled $Ftn^{(neg)}$ with (A) Gua condition, (B) pH 2 condition after 1st SEC and (C) pH 2 condition after 2nd SEC was crystallized with untreated $Ftn^{(pos)}$ to obtain large octahedral crystals. Scale bars are 200 µm.

The aim of this work is the creation of binary nanoparticle superlattices. Thus, the crystallization of both protein containers after the dis- and reassembly approach is crucial. This was performed with both protein containers treated with the Gua condition. Octahedral crystals were obtained (Figure 5.23) as for the untreated samples with smaller crystal size. The dis- and reassembly of the protein container harms the crystal growth slightly, but the crystal size is still adequate for the purpose of this work.

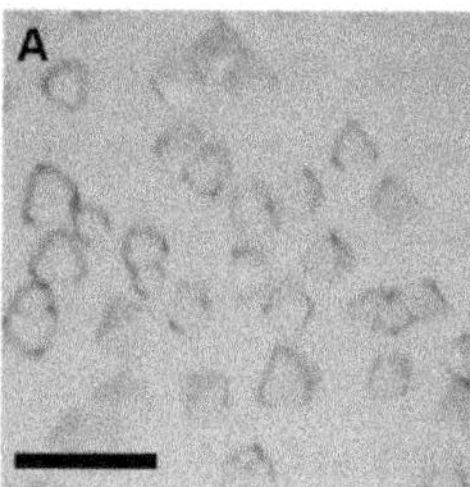

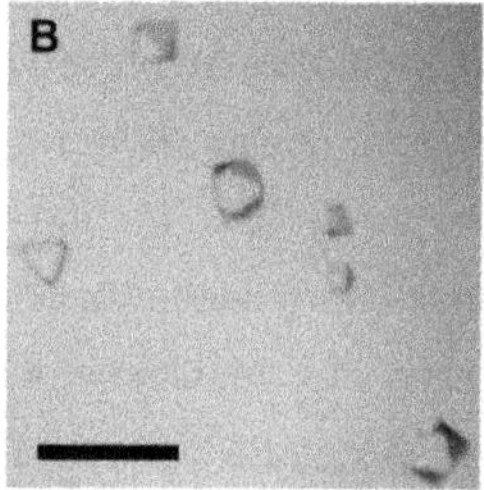

Figure 5.23: Crystallization of reassembled $Ftn^{(pos)}$ and $Ftn^{(neg)}$. (A) Crystallization of untreated protein containers. (B) Crystallization of both protein containers after dis- and reassembly in the Gua condition. Scale bars are 200 µm.

In summary, the dis- and reassembly in guanidinium is successful for both protein containers. In addition, the crystallization of one protein container or both protein containers after the dis- and reassembly led to large octahedral crystals. The dis- and reassembly of $Ftn^{(pos)}$ in acidic condition was not possible. However, for $Ftn^{(neg)}$ the protein containers were reformed again, but further purification is needed for the successful crystallization of the sample.

5.1.4 AuNP encapsulation

The synthesis of AuNP with a size of 3.4 ± 0.4 nm and the dis- and reassembly of both protein containers were successfully performed. In the last step, both results will be combined to encapsulate the synthesized AuNPs inside the cavity of both protein containers (Figure 5.24).

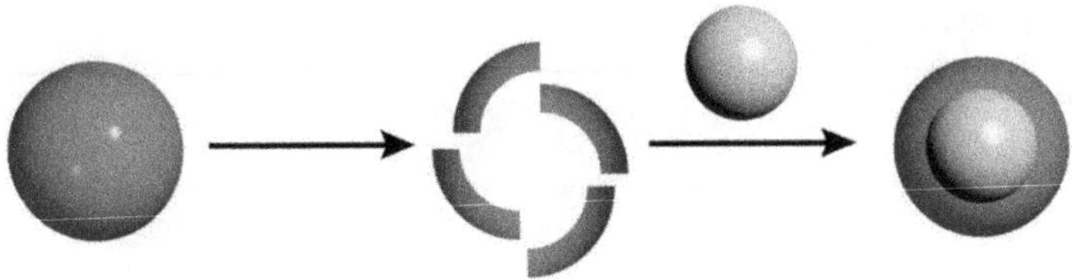

Figure 5.24: Encapsulation of AuNP inside the protein container cavity. After disassembly of the protein container, AuNPs are added to be encapsulated inside the protein container cavity during the reassembly process.

The influence of the AuNP ligand shell and the salt concentration during the reassembly for the encapsulation were studied. The last step was the optimization of the encapsulation efficiency to ensure a sample loaded completely with AuNPs and the upscaling of the sample to obtain sufficient sample amounts for crystallization experiments.

5.1.4.1 Influence of the ligand shell

In a first step, the influence of the ligand shell for the encapsulation was investigated. As discussed above, the characterization of the ligand shells showed a different stability for $AuC11^{+}$, AuC2 and AuC2mix due to the difference in surface charge. For a successful encapsulation of AuNPs, two factors are important: (I) Interaction of the AuNP's ligand shell with the inner surface of the protein container to ensure efficient encapsulation. (II) Colloidal stability of the AuNPs during the reassembly of the protein container to guarantee the presence of AuNPs during the encapsulation. The colloidal stability is also influenced by the salt concentration, which will be optimized later. For comparison, the concentration of sodium chloride was adjusted to 100 mM. Small amounts of $Ftn^{(neg)}$ were used and the encapsulation of the AuNPs was performed with the Gua and pH 2 condition. To investigate the influence of the ligand shell on the nanoparticle size of the encapsulated AuNPs, AuNP-Leff were used because of the larger size distribution. In this way, the upper size limit for encapsulation can be determined. Characterization was performed by TEM observing successful encapsulation and measuring the size of the encapsulated AuNPs. Experiments with Ftn$^{(pos)}$ were not performed because the focus was laid on the AuNP ligand shell system. Optimization of the encapsulation efficiency was performed mainly by variation of the salt concentration and ratio of AuNP to protein container later for each protein container variant separately. As mentioned in chapter 5.1.1, the AuNP should exceed a size of 3 nm to maintain a visible plasmon absorbance maximum, which is important for application as optical material.

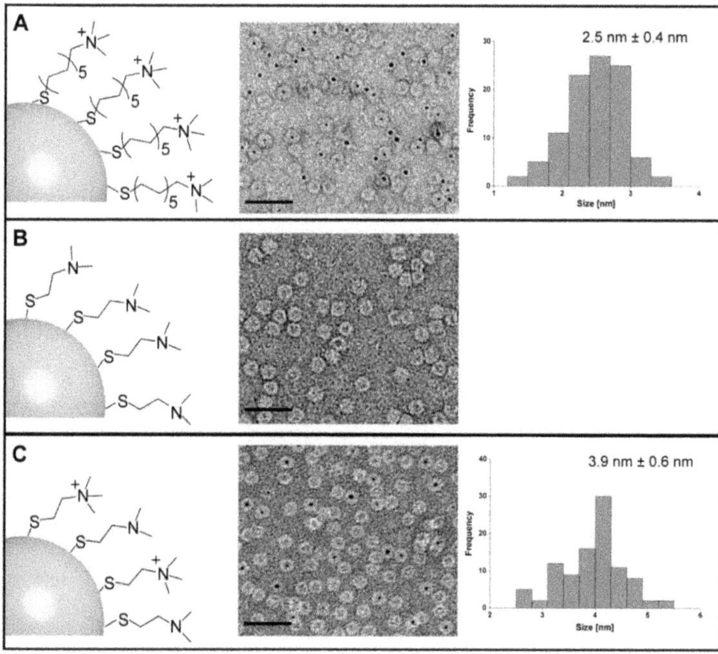

Figure 5.25: Encapsulation of different AuNPs. TEM images and size distributions of the encapsulation of (A) AuC11$^+$, (B) AuC2 and (C) AuC2mix. Scale bars are 50 nm.

The encapsulation of AuC11$^+$ and AuC2mix was successfully performed inside Ftn$^{(neg)}$ with the Gua condition. Interestingly, AuC11$^+$ was found both in the protein container cavity and outside of the protein container in solution (Figure 5.25A). Importantly, AuC2mix was completely encapsulated into the protein container cavity (Figure 5.25C). The C11$^+$ ligand is larger in size than the C2 and C2$^+$ ligand, which results in a larger total nanoparticle size. During the reassembly, a size exclusion effect takes place due to the size limitation of the protein container cavity so that larger nanoparticles were not encapsulated. As proof, the size of the encapsulated AuNP was determined. A size of 3.9 nm ± 0.6 nm was measured for AuC2mix and 2.5 nm ± 0.4 nm for AuC11^{+}. As excepted, the nanoparticle core size of encapsulated AuC11^{+} is smaller than for AuC2mix due to the longer ligand length of C11^{+}. Interestingly, if the ligand length is added to the nanoparticle size, the contrary result is observed. The AuC11^{+} has a total size of 5.6 nm and AuC2mix of 5.2 nm. The difference in total size explains why free AuNP were found for AuC11^{+} and not for AuC2mix. The upper size limit suitable for encapsulation was not reached for AuC2mix. Thus, an encapsulation of even larger AuNPs stabilized with the C2mix ligand shell is feasible. Compared with the metal oxide nanoparticle synthesized *in situ* inside the protein container cavity,[40] the AuNPs with ligands have a similar size. A size of 5.2 nm to 5.6 nm was measured, depending on the type of metal oxide nanoparticle. Furthermore, based on an inner cavity size of 6 nm, the cavity is 63% or 81% filled with AuC2mix or AuC11^{+}, respectively. However, no AuC2 was encapsulated or found freely outside the protein containers (Figure 5.25B). All AuC2 precipitated during the reassembly of the protein container. The amount of reformed protein containers is similar to the other two AuNPs. Thus, AuC2 did not agglomerate with the protein container during the reassembly process. The stability tests showed a lower colloidal stability of AuC2 in 0.1 M sodium chloride compared with AuC2mix and AuC11^{+}. The protein container fragments were not able to stabilize the AuC2 before it precipitated.

The encapsulation of AuNPs into Ftn$^{(neg)}$ with the pH 2 condition showed similar results compared with the encapsulation using the Gua conditions, although with some differences. Again, the encapsulation of AuC2 was not successful (Figure 5.26B) and no AuC2 was found at all in TEM. Interestingly, although the same amount of protein was used for all three experiments, the amount of reformed protein container was the highest for the AuC2 sample. Probably, an agglomeration of AuC2 occurred, resulting in precipitation of AuC2 without influencing the protein container reformation. Importantly, the encapsulation of AuC2mix and AuC11^{+} was successful for both samples. Similar to the Gua condition, a size exclusion effect due to the limited space in the protein container cavity occurred for AuC11^{+} (Figure 5.26A). Larger AuNP were not encapsulated and remained freely in solution, as observable in TEM. Thus, the sample contained empty and nanoparticle-loaded protein containers. Additionally, few AuNPs were attached to the negatively charged outer surface of the protein container due to electrostatic interactions. Interestingly, the encapsulation of AuC2mix (Figure 5.26C) led to only nanoparticle filled protein containers. Moreover, free AuNP were still in solution. Compared with the Gua condition, the nanoparticles showed an increased stability during the encapsulation with the pH 2 condition. Although the same amount of AuNPs was used for the encapsulation in Gua and pH 2, more AuNPs are visible in the TEM images indicating a higher stability of the AuNPs in the pH 2 condition compared with the Gua condition.

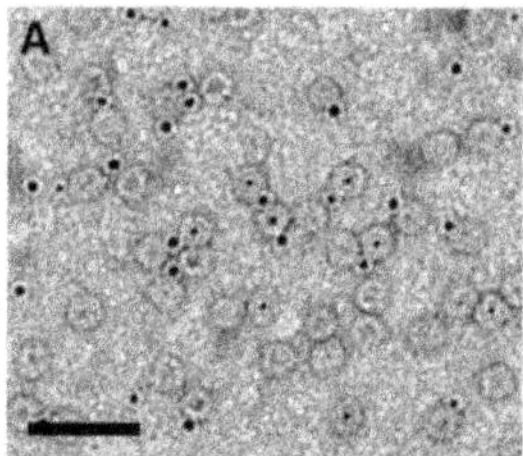

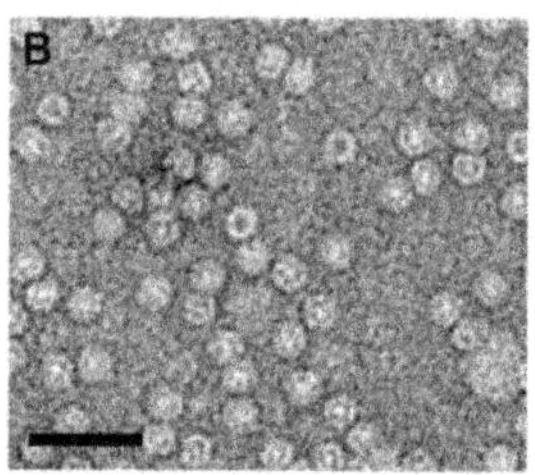

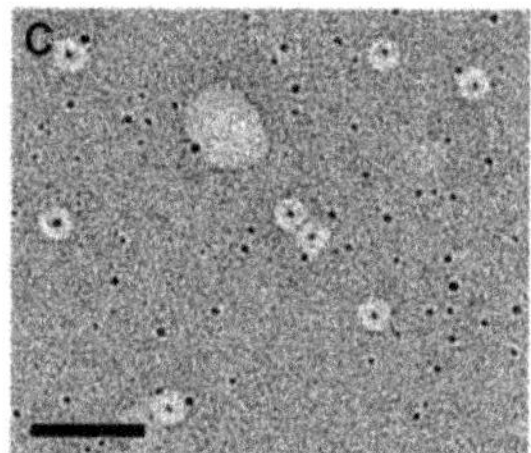

Figure 5.26: Encapsulation of different AuNPs with the pH 2 condition. TEM images of the encapsulation of (A) AuC11$^+$ (B) AuC2 and (C) AuC2mix. Scale bars are 50 nm.

On the other hand, only few reformed protein containers were found. In contrast to AuC11$^+$, the amount of reformed protein containers decreased drastically. The only difference between both samples is the amount of nanoparticle-loaded protein containers. For TEM measurements, the samples were concentrated using a centrifugal filter unit. During this process, the concentration of protein containers as well as AuNPs was increased. The negatively charged outer surface of Ftn$^{(neg)}$ can interact with the positively charged AuNPs leading to a precipitation of both components. This effect explains why the amount of protein containers was the highest for AuC2. Here, the AuNPs precipitated before concentrating the sample. In addition to coprecipitation of protein containers and AuNPs, the AuNP-loaded protein containers showed a lower stability compared with empty protein containers. If the empty and nanoparticle-loaded protein containers had the same stability, the same number of protein containers should be observed in the TEM images of both samples. Instead, more reformed protein containers are observed for AuC11$^+$.

As a result, only AuC2mix was used for further investigations. AuC2 was not able to be encapsulated inside the protein container cavity. Furthermore, the nanoparticle size of encapsulated AuNPs for the AuC11$^+$ fell below the critical nanoparticle size of 3 nm for the desired properties. Moreover, important factors for upscaling were determined that need to be considered in later experiments: Free AuNPs can interact with Ftn$^{(neg)}$ and can lead to a precipitation of the protein containers. The stability of AuNP-loaded protein containers is much lower than for empty protein containers.

5.1.4.2 Influence of the ionic strength

For the encapsulation of AuC2mix, the salt concentration plays a crucial role. The strength of the interaction between the positively charged AuNP and the negatively charged inner surface of the protein container can be adjusted by the variation of the salt concentration in solution during the encapsulation process. On the one hand, a high salt concentration will shield the charges of both components and no encapsulation of AuNP will take place at all. On the other hand, a low salt concentration will lead to a strong non-specific interaction of both components and precipitation of the sample. The aim is to find the best salt concentration to ensure a strong binding strength for an enhanced encapsulation efficiency without precipitation of the sample due to unwanted unspecific interactions, e.g. with the outer surface of the container. The salt screening had to be separately performed for both dis- and reassembly conditions as well as both protein containers. The Gua and pH 2 condition exhibit different ionic strength after the dis- and reassembly procedure. Both protein containers differ in their outer surface charge. The positively charged AuNPs can interact with the negatively charged inner as well as outer surface of $Ftn^{(neg)}$. However, in $Ftn^{(pos)}$ only the inner surface is negatively charged. The experimental procedures were the same as discussed above with a minor change. AuNP-Peng was used because of the higher suitability for encapsulation (see chapter 5.1.1.1). For each trial, the same amount of $Ftn^{(pos)}$ or $Ftn^{(neg)}$ was disassembled and AuNPs were added in aqueous solution, containing different salt concentrations. The analysis was performed by TEM, counting the number of empty and AuNP-filled protein containers. Figure 5.27A shows the number of filled containers for both protein containers with the Gua condition in different sodium chloride concentrations. It must be mentioned that the values are normalized against the highest value. The first aim was to find the best sodium chloride concentration for encapsulation. Further optimization will be done regarding the ratio of AuNPs to protein containers to obtain a sample with all protein containers filled with AuNP. Due to the different yields of reformed protein containers, it must be adjusted separately for each protein container. Interestingly, the same trend was visible for both protein containers. The best encapsulation efficiency was obtained at 200 mM sodium chloride. As expected, a higher sodium chloride concentration decreased the encapsulation efficiency drastically due to the enhanced shielding of the charges by the higher ionic strength. At 200 mM sodium chloride concentration, almost all $Ftn^{(pos)}$ were filled with AuNP ($AuFtn^{(pos)}$) and a lot of free AuNPs were in solution (Figure 5.27B). By increasing the concentration to 500 mM (Figure 5.27C) the amount of protein containers as well as AuNP decreased extremely. The higher salt concentration led to precipitation of the free AuNPs, possibly also precipitating the protein containers. In contrast, the amount of Ftn$^{(neg)}$ remained constant. At 200 mM (Figure 5.27D) and 500 mM (Figure 5.27E) the free AuNPs are attached to the negatively charged outer surface. The attachment of the AuNP to the protein container stabilized the AuNP additionally, which prevented the precipitation. Therefore, the lower encapsulation efficiency at higher ionic strength is mainly caused by the stronger shielding of the different charges and not by precipitation of the AuNPs.

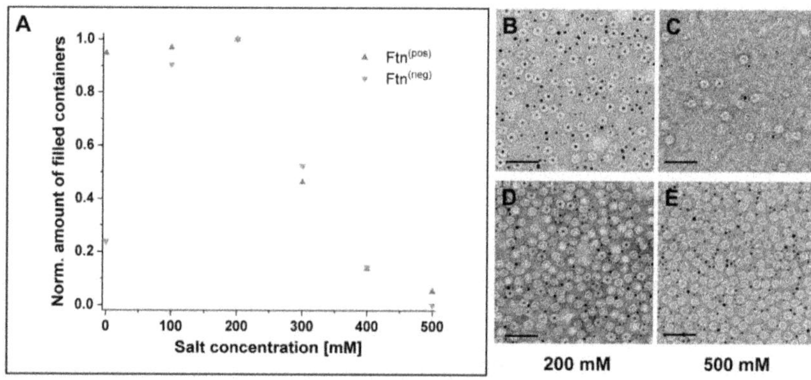

Figure 5.27: Salt screening for the encapsulation of AuC2mix with the Gua condition. (A) Normalized amount of filled protein containers for Ftn$^{(pos)}$ (blue) and Ftn$^{(neg)}$ (red). The highest encapsulation efficiency was obtained at 200 mM sodium chloride. TEM images of the encapsulation of AuC2mix in Ftn$^{(pos)}$ at 200 mM (B) and 500 mM (C) sodium chloride concentration as well as in Ftn$^{(neg)}$ at 200 mM (D) and 500 mM (E). Scale bars are 50 nm.

The procedure was repeated for the encapsulation of AuC2mix in Ftn$^{(neg)}$ with the pH 2 condition. In contrast to the Gua condition, the ionic strength of the pH 2 condition should be lower. After reassembly in the Gua condition, 0.5 M Gua still remained, while in the pH 2 condition a sodium chloride concentration of 0.05 M was obtained. Moreover, the pH value in the reassembly solutions differed from the Gua condition and so the total surface charge of AuC2mix. Accordingly, the encapsulation at 200 mM sodium chloride (Figure 5.28A) led to AuNP-filled Ftn$^{(neg)}$ (AuFtn$^{(neg)}$), but not to the highest encapsulation efficiency. By increasing the sodium chloride concentration to 400 mM the best encapsulation efficiency was obtained (Figure 5.28B). Interestingly, a lower ionic strength is required for the best encapsulation efficiency compared to the Gua condition (0.54 mM vs. 0.70 mM). This effect can be explained by the different pH values of the reassembly solutions, resulting in slightly different charged AuNPs during the encapsulation procedure. Thus, the strength of the electrostatic interaction between the negatively charged inner surface of the protein container and the positively charged AuNPs differs for both encapsulation conditions. Similar to the Gua condition, a higher ionic strength of the solution prevented the encapsulation of AuC2mix in Ftn$^{(neg)}$ (Figure 5.28C) due to the stronger shielding of the charged components.

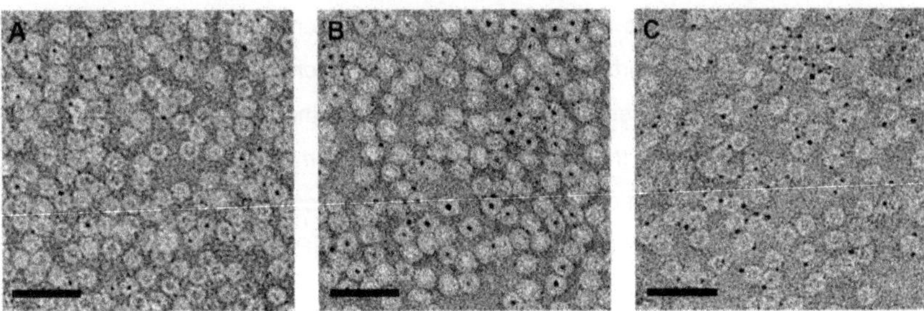

Figure 5.28: Salt screening for the encapsulation of AuC2mix with the pH 2 condition. TEM images of the encapsulation of AuC2mix in Ftn$^{(neg)}$ at 200 mM (A), 400 mM (B) and 600 mM (C) sodium chloride concentration. The highest encapsulation efficiency was obtained at 400 mM sodium chloride. Scale bars are 50 nm.

As a result, the nanoparticle encapsulation efficiency can be easily influenced by variation of the sodium chloride concentration. By increasing the sodium chloride concentration, the shielding of the opposite charges was enhanced leading to decreased interaction between the components and thus reduced encapsulation. The best encapsulation efficiency was obtained at 200 mM for the Gua condition and 400 mM for the pH 2 condition.

5.1.4.3 Purification and optimization

After the best ligand shell and salt concentration were determined for the encapsulation of AuNP, two main problems occurred. On the one hand, it was not possible to obtain a sample with all protein containers filled with AuNP. Empty protein containers were still present next to nanoparticle-filled protein containers. On the other hand, not all AuNPs were encapsulated and remained freely in solution. The preparation of the samples for TEM showed a precipitation of protein containers in the presence of free AuNPs. Moreover, these free AuNP would be troublesome for crystallization of the samples. In a first step, the separation of free AuNPs from protein containers was investigated. The purification of the sample can be performed with size exclusion chromatography or ion exchange chromatography (IEC). Afterwards tackling this challenge, the amount of filled protein containers will be optimized in further experiments (see below).

5.1.4.3.1 Purification of AuFtn$^{(pos)}$

The separation of Ftn$^{(pos)}$ from free AuC2mix can be performed using the difference between the two species regarding size or charge. In detail, AuC2mix have a size of approximately 6 nm and Ftn$^{(pos)}$ of 12 nm. Both components are positively charged. AuC2mix had a ζ-pot. of 27 mV and Ftn$^{(pos)}$ of 19.2 mV. Therefore, the purification was investigated by SEC due to the larger size difference compared to the charge difference. To this end, the absorbance was measured additionally at 520 nm to detect the AuNP, both the free particles and encapsulated within the ferritin containers (co-elution). For the experiments, the encapsulation was performed with 2 mg Ftn$^{(pos)}$ and a 1:1 ratio of AuNP to protein container. Moreover, a salt concentration of 200 mM was used. Figure 5.29 showed the SEC chromatogram of AuFtn$^{(pos)}$.

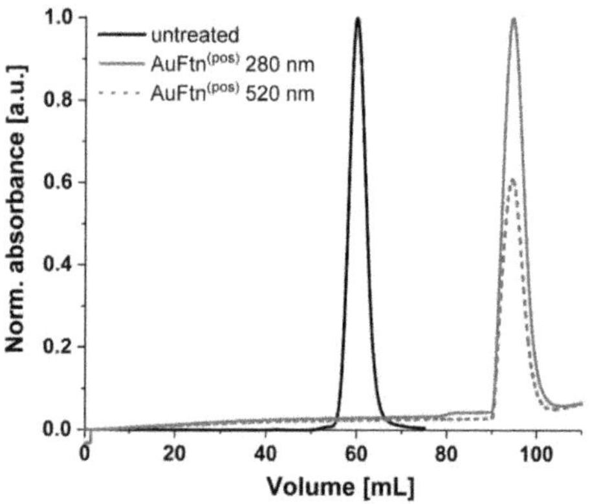

Figure 5.29: SEC chromatogram of AuFtn$^{(pos)}$. Size exclusion chromatogram of Ftn$^{(pos)}$ after encapsulation of AuC2mix. The untreated Ftn$^{(pos)}$ in black and AuFtn$^{(pos)}$ in blue.

The purification of AuFtn$^{(pos)}$ led to one sharp peak at an elution volume of 95 mL. Unfortunately, the peak is shifted to higher values compared with the untreated protein container, which indicated a smaller size. Whereas the peak showed a high signal at 520 nm, it should only contain AuNPs due to the smaller total size of the nanoparticles. TEM images of the fractions confirmed the presence of only AuNPs (Appendix, Figure 8.7). No signal at 60 mL for the protein container at all was measured. During the purification, AuFtn$^{(pos)}$ and Ftn$^{(pos)}$ precipitated on the column. Therefore, as an alternative, the purification via IEC was investigated. For the purification by IEC, it was crucial if the small charge difference is large enough for an adequate separation of both components. However, the advantage of IEC is that it is independent of the sample volume: For sample injection in SEC, the sample had to be concentrated to at least 2 mL using a centrifugal filter unit. By this process, a certain amount of sample was lost due to precipitation on the filter. For IEC, no specific volume was required for sample injection. The sample was loaded directly onto the column with a solution of 0.1 M sodium chloride at pH 7.5. Elution was performed with a gradient to 1.5 M sodium chloride at pH 7.5. The chromatogram showed two very well separated peaks with high absorbance intensity at 520 nm (Figure 5.30A). Due to the lower positive surface charge (see above), $Ftn^{(pos)}$ should bind weaker and elute at smaller volumes compared with the AuNP. TEM images of the first (Figure 5.30B) and second peak (Figure 5.30C) confirmed this assumption. The first peak contained $Ftn^{(pos)}$ and $AuFtn^{(pos)}$. The second peak consisted mainly of free AuNP and a few $AuFtn^{(pos)}$.

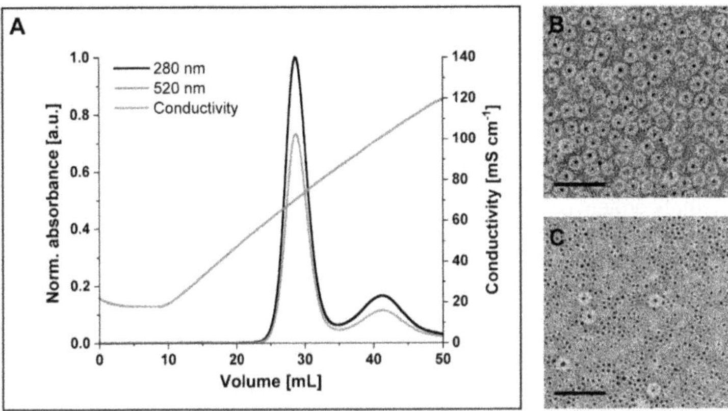

Figure 5.30: Purification of $AuFtn^{(pos)}$ by IEC. (A) Cation exchange chromatogram of $Ftn^{(pos)}$ after encapsulation of AuC2mix. Absorbance at 280 nm is shown in black and at 520 nm in red. Conductivity is shown in green. TEM images of the first (B) and second peak (C) of the chromatogram. Scale bars are 50 nm.

The last step was the improvement of the encapsulation efficiency, thus having mostly filled $Ftn^{(pos)}$ and ideally no empty protein containers. The separation of both components after encapsulation by ultracentrifugation with a sucrose gradient was not successful. $AuFtn^{(pos)}$ precipitated during the centrifugation. However, the number of $AuFtn^{(pos)}$ can easily be increased to 98 % by changing the ratio of AuNP to $Ftn^{(pos)}$ to 2:1 (Figure 5.31). Free AuNP are separated from the filled protein containers by IEC.

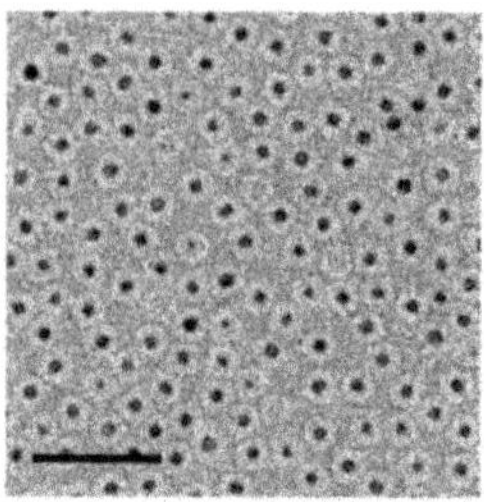

Figure 5.31: Optimization of AuNP loading in AuFtn$^{(pos)}$. TEM image of AuFtn$^{(pos)}$ with 98 % filled Ftn$^{(pos)}$. Scale bar is 50 nm.

The purity of the sample and the size of the protein containers were determined by DLS and TEM (Table 5.8). For AuFtn$^{(pos)}$ a hydrodynamic radius of 15.94 nm was obtained. Compared with untreated Ftn$^{(pos)}$ the hydrodynamic radius increased. Additionally, the PDI is higher but still in the range of a rather monodisperse sample. The size determined by TEM is the same for AuFtn$^{(pos)}$ and untreated Ftn$^{(pos)}$. Thus, the AuNPs did not alter the protein container size during the encapsulation process. Moreover, the ζ-pot. was measured and as expected, a similar value for AuFtn$^{(pos)}$ was obtained as for the empty protein container.

Table 5.8: DLS, TEM and ζ-pot. data for AuFtn$^{(pos)}$, Ftn$^{(pos)}$ and AuNP.

Sample	d_{DLS} [nm]	PDI	d_{TEM} [nm]	ζ-pot. [mV]
Ftn$^{(pos)}$	14.57	0.030	14.6 ± 1.7	19.2 ± 1.7
AuFtn$^{(pos)}$	15.94	0.117	14.5 ± 1.3	17.6 ± 0.4
AuNP			3.4 ± 0.4	27.0 ± 1.8
AuNP in AuFtn$^{(pos)}$			3.4 ± 0.4	

In summary, purification by IEC was performed to separate free AuNP from protein containers. By increasing the AuNP to protein container ratio to 2:1, a sample with 98 % AuNP filled protein containers was obtained. The size of the protein container as well as the AuNP remained unchanged during the encapsulation process.

5.1.4.3.2 Purification of AuFtn$^{(neg)}$

The purification of AuFtn$^{(neg)}$ can be performed similarly to AuFtn$^{(pos)}$ with SEC or IEC. AuNPs can be encapsulated with the Gua condition as well as pH 2 condition (see above). First, the encapsulation with the Gua condition was investigated. Here, the results of the AuFtn$^{(pos)}$ purification were used. The purification was directly performed with IEC and the AuNP to Ftn$^{(neg)}$ ratio was increased to 2:1. The opposite charge of the AuNP and Ftn$^{(neg)}$ will make the separation easier. By using an anionic exchange column, the free AuNP will not bind to the column and run through the column. Meanwhile AuFtn$^{(neg)}$ will bind to the column. Subsequently, AuFtn$^{(neg)}$ will be collected by increasing the salt concentration of the running buffer. The sample was loaded with 0.1 M sodium chloride at pH 7.5 and a gradient was run to 1.0 M sodium chloride at pH 7.5. The IEC chromatogram showed the expected results (Figure 5.32A). The broad peak at the beginning between 0 mL and 35 mL belonged to the free AuNP, which did not bind to the column. The second sharp peak at an elution volume of 62 mL should contain the AuNP-filled protein containers. The TEM image of this peak (Figure 5.32B) confirmed the assumption.

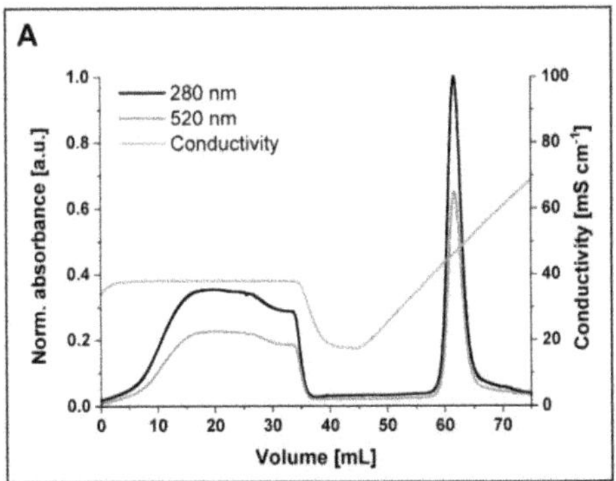

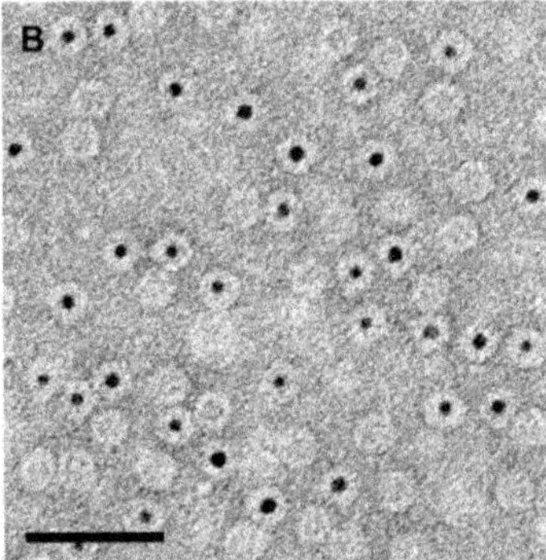

Figure 5.32: Purification of AuFtn$^{(neg)}$ by IEC. (A) Anion exchange chromatogram of Ftn$^{(neg)}$ after encapsulation of AuC2mix. Absorbance at 280 nm is shown in black and at 520 nm in red. Conductivity is shown in green. (B) TEM image of the sharp peak at 62 mL. Scale bar is 50 nm.

The sample contained AuFtn$^{(neg)}$ and Ftn$^{(neg)}$. In contrast to AuFtn$^{(pos)}$, the amount of filled protein containers compared to empty protein containers was still low. The yield after the dis- and reassembly procedure is for Ftn$^{(neg)}$ three times higher than for Ftn$^{(pos)}$. Thus, a higher AuNP to protein container ratio was needed, as more Ftn$^{(neg)}$ protein containers survive the dis- and reassembly process. The ratio of AuNP to protein containers was increased to 4.5:1 to get 92 % AuNP-filled Ftn$^{(neg)}$ (Figure 5.33). A further increase of AuNP led to a lower encapsulation rate.

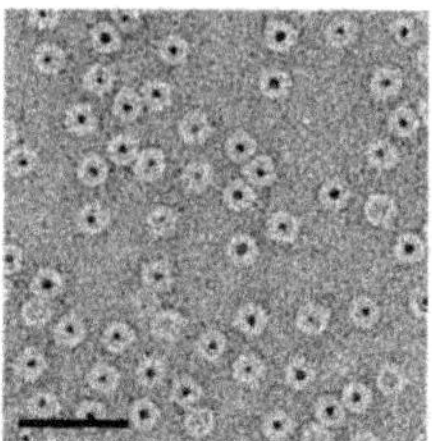

Figure 5.33: Optimization of AuNP loading in AuFtn$^{(neg)}$ with Gua condition. TEM image of AuFtn$^{(neg)}$ with 92 % filled Ftn$^{(neg)}$. Scale bar is 50 nm.

The purity and size of the samples were measured by DLS and TEM (Table 5.9). AuFtn$^{(neg)}$ had a higher hydrodynamic radius and PDI compared with the untreated Ftn$^{(neg)}$. The values indicated a lower purity of the sample, which could be problematic for the crystallization of the sample. Further purification is mandatory. However, the size determined by TEM is in the expected size range (14.6 nm). Moreover, the ζ-pot. was measured and as expected a negative charge of -20.5 mV was obtained.

Table 5.9: DLS, TEM and ζ-pot. data for AuFtn$^{(neg)}$ and Ftn$^{(neg)}$.

Sample	d_{DLS} [nm]	PDI	d_{TEM} [nm]	ζ-pot. [mV]
Ftn$^{(neg)}$	14.99	0.078	15.0 ± 1.0	-23.7 ± 1.3
AuFtn$^{(neg)}$ Gua	21.33	0.354	14.6 ± 1.2	-20.5 ± 0.7

The higher hydrodynamic radius and PDI for the AuFtn$^{(neg)}$ can be caused by aggregates, which were formed during the encapsulation. A straight-forward way to purify the sample would be a SEC. The SEC of AuFtn$^{(pos)}$ led to precipitation of the sample on the column. Presumably, the precipitation was caused by the free AuNPs. Here, for AuFtn$^{(neg)}$, after the purification by IEC, the sample was free from non-encapsulated AuNPs. The concentration of the sample to a volume of 2 mL was possible without precipitation of sample on the filter of the centrifugal filter unit. SEC was performed with the usual running buffer of the untreated protein container (Figure 5.34). The SEC chromatogram showed one sharp peak at the elution volume of the untreated Ftn$^{(neg)}$. Moreover, increased purity was confirmed by DLS measurements. The hydrodynamic radius and PDI decreased to 14.60 nm respectively 0.127. Thus, similar values as for untreated Ftn$^{(neg)}$ were obtained.

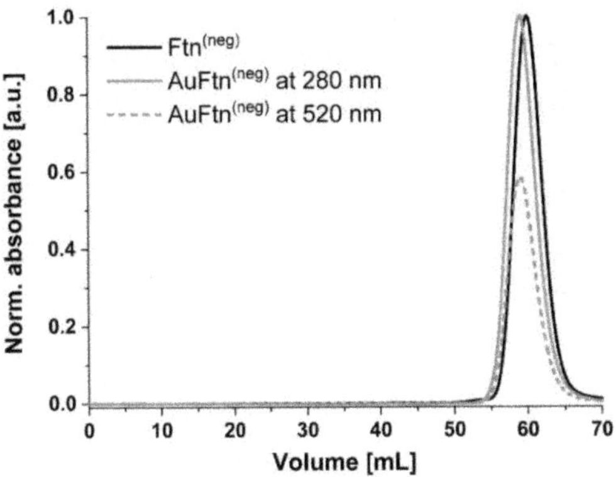

Figure 5.34: Additional purification of AuFtn$^{(neg)}$ by SEC. Additional SEC of AuFtn$^{(neg)}$ after purification with IEC. The untreated Ftn$^{(neg)}$ in black and AuFtn$^{(neg)}$ in red. The absorbance was measured at 280 nm (line) and at 520 nm (dots).

Subsequently, the purification of the encapsulation with the pH 2 condition was investigated. First, a purification by SEC was performed. The sample preparation was the same as for the Gua condition. 2 mg Ftn$^{(neg)}$ and an AuNP to Ftn$^{(neg)}$ ratio of 2:1 was used. The concentration of the sample to a volume of 2 mL with the centrifugal filter unit led to no precipitation of sample on the filter. In contrast to the Gua condition, the SEC showed a different behavior (Figure 5.35). AuFtn$^{(neg)}$ did not precipitate on the column during the purification. A sharp peak with a small shoulder was obtained shifted slightly to lower elution volumes compared with the untreated protein container. The measured absorbance at 520 nm indicated the presence of AuNPs. From the results of the encapsulation in Ftn$^{(pos)}$, free AuNP were expected at higher elution volumes around 95 mL. Thus, the signal at 60 mL elution volume should originate from AuNPs encapsulated in protein containers. Subsequently, a second SEC purification was performed. The investigation of the dis- and reassembly of Ftn$^{(neg)}$ in the pH 2 condition revealed the necessity of further purification for the empty sample. Without this additional purification the crystallization led to ugly crystals with many defects due to the impurity of the sample. In the second SEC, one single sharp peak at the same elution volume as the untreated Ftn$^{(neg)}$ was obtained.

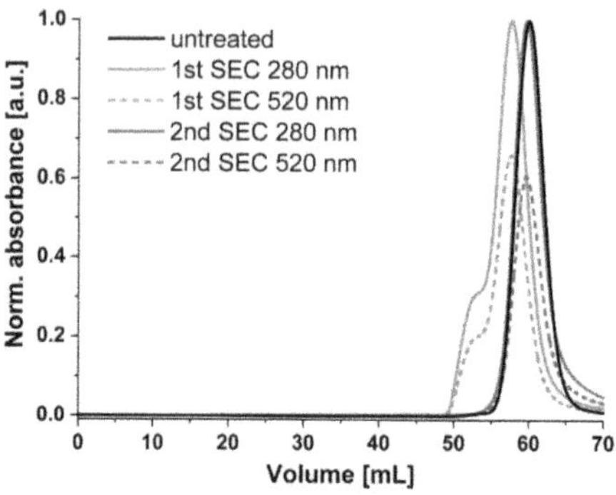

Figure 5.35: SEC chromatogram of AuFtn$^{(neg)}$. SEC chromatogram of Ftn$^{(neg)}$ after encapsulation of AuC2mix with pH 2 condition. The untreated Ftn$^{(neg)}$ in black, 1st SEC of AuFtn$^{(neg)}$ in red and 2nd SEC in blue. The absorbance was measured at 280 nm (line) and at 520 nm (dots).

Further investigation was performed by TEM (Figure 5.36). The sample contained 88 % of AuNP filled protein containers. In contrast to the encapsulation with the Gua condition, less AuNP had to be used for the encapsulation to obtain an adequate loading of Ftn$^{(neg)}$. Therefore, the encapsulation efficiency is higher in the pH 2 condition. Moreover, AuFtn$^{(neg)}$ treated with the pH 2 condition showed a similar hydrodynamic radius (14.63 nm) and PDI (0.132) as untreated Ftn$^{(neg)}$.

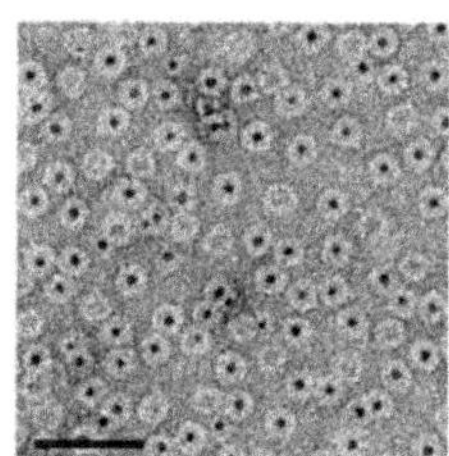

Figure 5.36: Optimization of AuNP loading in AuFtn$^{(neg)}$ with pH 2 condition. TEM image of AuFtn$^{(neg)}$ with 88 % filled Ftn$^{(neg)}$. Scale bar is 50 nm.

In total, the separation of protein containers from free AuNP was performed by IEC for the Gua condition and by SEC for the pH 2 condition. Moreover, after removal of the non-encapsulated AuNPs, a further purification with SEC of the Gua sample was feasible without losing the sample on the column. On the one hand, the pH 2 condition showed a higher encapsulation efficiency, on the other hand a slightly lower amount of AuFtn$^{(neg)}$ compared with the Gua condition. Finally, AuNP-loaded protein containers were obtained as building blocks ready for the assembly into the highly ordered material.

5.1.5 AgNP encapsulation

For the construction of optical nanomaterials based on gold and silver nanoparticles, the next step was the encapsulation of AgNPs into the protein container cavity. Some important results from the AuNP encapsulation will be adopted for the AgNP encapsulation: (I) The averaged size of the AgNPs was 3.1 nm (see 5.1.2.1). The encapsulation of AuNPs showed discrete size limitations for each ligand shell. On the one hand, $AuC11^+$ nanoparticles with a size of 2.5 nm were encapsulated, on the other hand AuC2mix with a size of 3.9 nm. Thus, only the C2mix ligand shell was utilized for the encapsulation. (II) The sodium chloride concentration had a large effect on the encapsulation efficiency. The best concentrations were adopted for the AgNP encapsulation. (III) Free nanoparticles influenced the stability of protein containers during the concentration with centrifugal unit filters. A separation of both components by IEC is required. The stability experiments with AgC2mix revealed a lower colloidal stability compared with AuNPs. Moreover, AgC2mix was highly unstable in 7 M Gua, which is used as disassembly condition for both protein containers. Encapsulation processes were directly investigated with larger batches of protein containers. The samples were purified with IEC. TEM measurements before and after IEC as well as after concentration of the sample were performed.

5.1.5.1 Encapsulation in $Ftn^{(pos)}$

The encapsulation into $Ftn^{(pos)}$ can only be performed in the Gua condition because the dis- and reassembly was only possible in the Gua condition (see 5.1.3.1). A ratio of protein container to AgNP of 1:1 was used. TEM imaging before IEC showed some empty protein containers and many free AgNP (Figure 5.37A) despite a low colloidal stability of AgC2mix in the Gua condition. No nanoparticle filled protein containers were found. Nevertheless, IEC of the sample was performed. The presence of AgNPs was monitored by absorbance at 420 nm. Interestingly, the chromatogram showed one discrete peak with an absorbance at 420 nm, which has a higher intensity than the absorbance at 280 nm (Figure 5.37B). The peak position was identical to the elution volume of $AuFtn^{(pos)}$. Thus, it was expected that sample with AgNP filled $Ftn^{(pos)}$ ($AgFtn^{(pos)}$) was obtained. TEM imaging of the sample revealed some black spots on the TEM grid, which can be assigned to AgNPs (Figure 5.37C). However, the round shell around the AgNPs is caused by positive staining of the ligand shell. Protein containers are larger in size and thus only free AgNPs were imaged. After concentration of the sample only empty Ftn^(pos)^ were found. The further treatment of the sample led to precipitation of the AgNPs (Figure 5.37D).

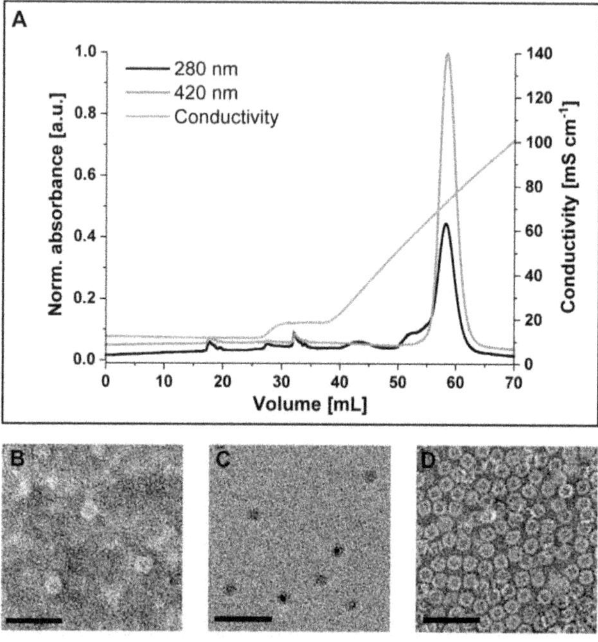

Figure 5.37: Purification of AgFtn(pos) by IEC. (A) Cation exchange chromatogram of Ftn(pos) after encapsulation of AgC2mix. Absorbance at 280 nm is shown in black and at 420 nm in red. Conductivity is shown in green. TEM images of the sample (B) before IEC, (C) after IEC and (D) after concentration of the sample. Scale bars are 50 nm.

In conclusion, the encapsulation of AgNPs into Ftn(pos) was not successful. To enhance the stability of the AgNPs, a change to the pH 2 condition would be mandatory. However, dis- and reassembly of Ftn(pos) in the acidic condition is not successful. At this point, synthesis of new AgNPs with an enhanced stability will be required so that no further encapsulation experiments were performed.

5.1.5.2 Encapsulation in Ftn(neg)

The dis- and reassembly of Ftn(neg) is possible with the Gua condition as well as the pH 2 condition. Due to the higher colloidal stability of AgC2mix in acidic pH values than in high guanidinium concentration, the pH 2 condition was favored. The procedure was the same as for the encapsulation in Ftn(pos). TEM image of the sample before IEC did not show encapsulated AgNP in Ftn(neg) (Figure 5.38B). Only empty protein containers were visible. Again, in the ion-exchange chromatogram (Figure 5.38A) one single discrete peak appeared with a higher absorbance at 420 nm than 280 nm. However, no AgNPs encapsulated in Ftn$^{(neg)}$ (AgFtn$^{(neg)}$) were found in the TEM image after IEC (Figure 5.38C) as well as after concentration of the sample (Figure 5.38D). Free AgC2mix was present explaining the high absorbance at 420 nm. Although no AgC2mix should bind to the column, it was eluted together with the empty protein container.

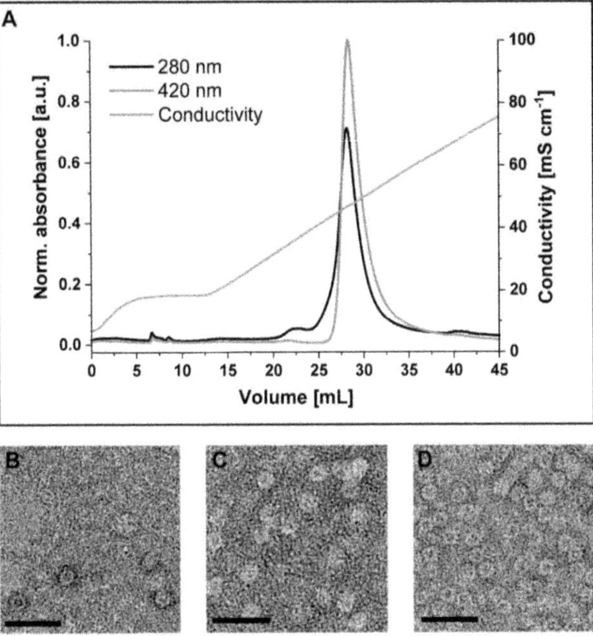

Figure 5.38: Purification of AgFtn$^{(neg)}$ by IEC. (A) Anion exchange chromatogram of Ftn$^{(neg)}$ after encapsulation of AgC2mix. Absorbance at 280 nm is shown in black and at 420 nm in red. Conductivity is shown in green. TEM images of the sample (B) before IEC, (C) after IEC and (D) after concentration of the sample. Scale bars are 50 nm.

In summary, the encapsulation in Ftn$^{(neg)}$ was not possible, although many AgC2mix survived the whole treatment. Due to the low colloidal stability of the AgNPs, an encapsulation was strongly hampered. To improve the encapsulation, an enhancement of the AgNP stability will be required. Therefore, a new ligand shell or a new nanoparticle synthesis must be found, matching the requirements for the AgNP.

5.2 Encapsulation of dye molecules

In this work, the goal was the assembly of AuNPs and AgNPs into a highly order material. Due to the difficulties with the AgNP encapsulation, an alternative building block was required. In previous studies, small molecules such as metal complexes,[172,257,258] drugs[173,179] or dyes[259–261] have been encapsulated in apoferritin by using dis- and reassembly approaches in acidic pH or by thermal diffusion through the protein container pores. The combination of plasmonic nanoparticles and dye molecules in a material could be interesting due to possible plasmon-exciton coupling.[262] Depending on the coupling strength a diverse range of effects can occur including fluorescence quenching/enhancement (weak coupling), Fano interference (intermediate coupling) and Rabi-splitting (strong coupling).[263] Rhodamine was chosen as dye because of its high fluorescence quantum yield and absorption maximum in the same wavelength range as the AuNPs. First, rhodamine 6G was encapsulated in $Ftn^{(pos)}$ by a thermal diffusion strategy. Importantly, rhodamine 6G is positively charged, which will improve the encapsulation efficiency. On the one hand, interactions between the dye and the outer surface are hampered due to the same charge and thus preventing the precipitation of the protein container. On the other hand, the diffusion through the pores is enhanced. The electrostatic field gradient of the pores can direct the positively charged dye to the channel entrances and guide it to the inside of the cavity. Moreover, the negatively charged inner surface enables an electrostatic binding of the dye to capture the dye inside the cavity. The protein container was incubated in pH 5 at 65°C while rhodamine 6G was added in more than 8000-fold excess. Non-encapsulated rhodamine 6G was separated by repeated washing with water in a centrifugal filter. The encapsulation of the dye in $Ftn^{(pos)}$ was investigated by SEC (Figure 5.39A). The absorption spectrum of rhodamine 6G (Appendix, Figure 8.8) shows a distinct absorption maximum at 530 nm. The absorption between 200 nm and 300 nm is very weak. Therefore, presence of the dye was monitored by absorbance at 530 nm, while the protein container was detected at 280 nm. The SEC chromatogram showed one broad peak. A high absorbance at 280 nm as well as 530 nm was measured for this peak. This observation indicates that both components were present. Compared with $Ftn^{(pos)}$, the peak is shifted to higher elution volumes. An attachment of dye molecules at the outer surface would result in a larger size and lower elution volume. Thus, the dye should be encapsulated inside $Ftn^{(pos)}$ ($RhFtn^{(pos)}$).

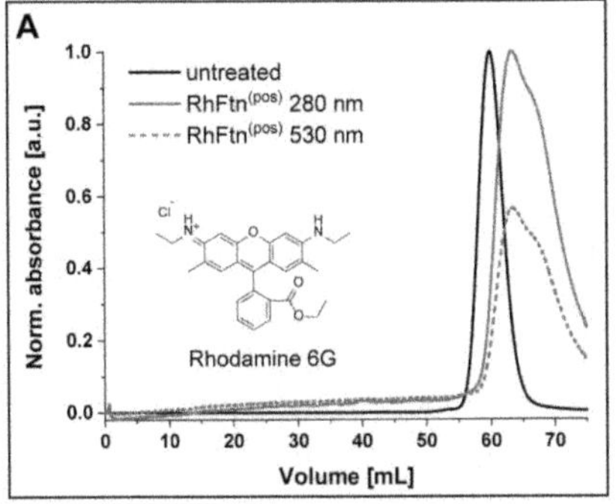

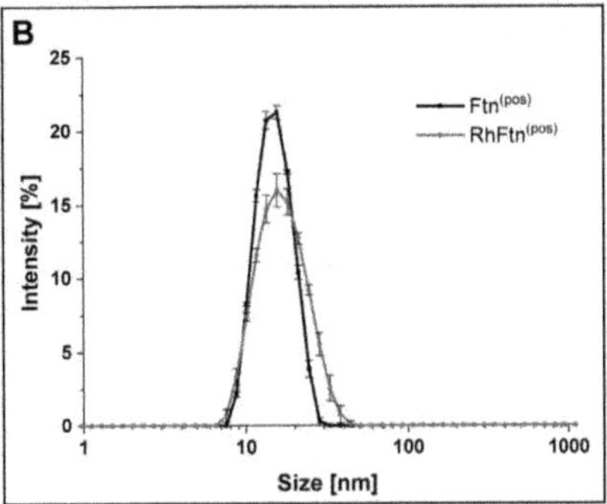

Figure 5.39: Encapsulation of rhodamine 6G in Ftn$^{(pos)}$. (A) Size exclusion chromatogram of RhFtn$^{(pos)}$. Absorbance of the protein was measured at 280 nm and of rhodamine 6G at 530 nm. The untreated protein container is displayed in black and RhFtn$^{(pos)}$ in blue. (B) Intensity-weighted size distribution of Ftn$^{(pos)}$ (black) and RhFtn$^{(pos)}$ (blue) as determined by DLS.

The sample was further investigated by DLS (Figure 5.39B). Comparison with Ftn$^{(pos)}$ revealed the same size, but with a broader size distribution. A hydrodynamic radius of 15.33 ± 0.23 nm was measured. Subsequently, the same dis- and reassembly approach using the pH 2 condition was performed as for the plasmonic nanoparticles to encapsulate rhodamine B in Ftn$^{(neg)}$. The dye was changed from rhodamine 6G to rhodamine B because the encapsulation of rhodamine 6G in Ftn$^{(neg)}$ led to precipitation of the protein container. The positively charged dye can interact electrostatically with the negatively inner and outer surface of the protein container. Presumably, the interaction of the dye with both surfaces prevented the reassembly of the protein subunits into the protein container. On the other hand, rhodamine B is zwitterionic, which lead to a weaker interaction with the outer and inner surface of the protein container. Thus, the weaker interaction of the dye with the protein subunits enables to reassemble into the protein container and simultaneously to capture dye molecules inside the cavity. For the encapsulation, a more than 10000-fold excess of rhodamine B was used. Again, the encapsulation was investigated by SEC (Figure 5.40A). The presence of the dye was monitored by absorbance at 550 nm. The encapsulation of dye in Ftn$^{(neg)}$ (RhFtn$^{(neg)}$) was successful. A distinct peak was obtained at the same elution volume as for empty Ftn$^{(neg)}$. Absorbance at 550 nm confirmed the presence of dye. Compared to RhFtn$^{(pos)}$ the loading of dye is lower due to a weaker absorbance. By DLS measurements (Figure 5.40B) a hydrodynamic radius of 15.09 ± 0.08 nm was obtained.

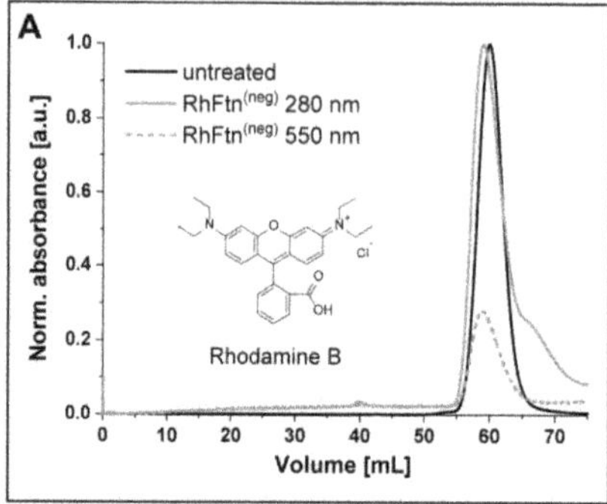

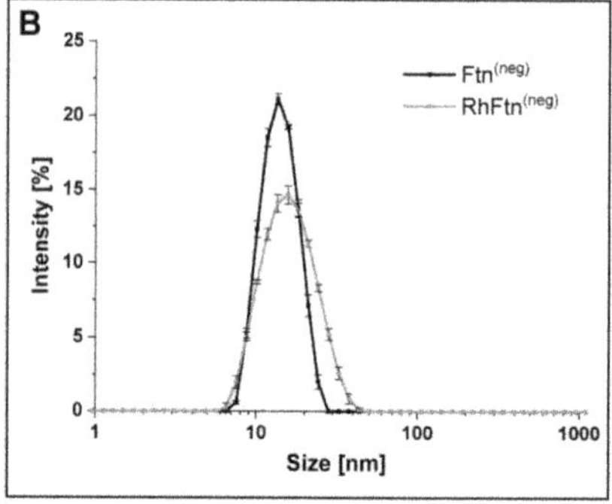

Figure 5.40: Encapsulation of rhodamine B in Ftn$^{(neg)}$. (A) Size exclusion chromatogram of RhFtn$^{(neg)}$. Absorbance of the protein was measured at 280 nm and of rhodamine B at 550 nm. The untreated protein container is displayed in black and RhFtn$^{(neg)}$ in red. (B) Intensity-weighted size distribution of Ftn$^{(neg)}$ (black) and RhFtn$^{(neg)}$ (red) as determined by DLS.

The amount of encapsulated dye molecules inside both protein container variants was established by Made Budiarta using UV-Vis spectroscopy (see 7.13.3). Similarly, for this sample, 2 rhodamine 6G molecules and 0.5 rhodamine B molecules per protein container were loaded in Ftn$^{(pos)}$ or Ftn$^{(neg)}$, respectively. A further increase of the dye to protein container ratio yielded more aggregated protein containers and thus a lower yield.

In total, the encapsulation of dye was successful for both protein containers. SEC and DLS measurements confirmed reformed protein containers. The presence of dye was proven with measured absorbance at the absorbance maximum of the dyes. However, the loading of dye was low for both protein variants.

5.3 Synthesis of metal oxide nanoparticles *in situ*

In addition to the encapsulation of pre-synthesized nanoparticles, another way to incorporate nanoparticles in the protein container cavity is the direct synthesis inside the cavity. For example, metal precursor and oxidation agent can enter the protein container cavity through pores where the nanoparticle can be formed. Hence, the protein container cavity will act as reaction vessel. In our group, the syntheses of metal oxide nanoparticles such as cerium oxide, cobalt oxide and iron oxide were optimized.[40] For this work, cerium oxide nanoparticles were synthesized inside Ftn$^{(pos)}$ (CeFtn$^{(pos)}$) as well as Ftn$^{(neg)}$ (CeFtn$^{(neg)}$) according to previous works. Both samples were purified with SEC (Figure 5.41A) and empty protein containers were separated from nanoparticle filled protein containers by sucrose gradient centrifugation (Appendix, Figure 8.9). A stained and unstained TEM image show protein containers and cerium oxide nanoparticles (Figure 5.41B and C).

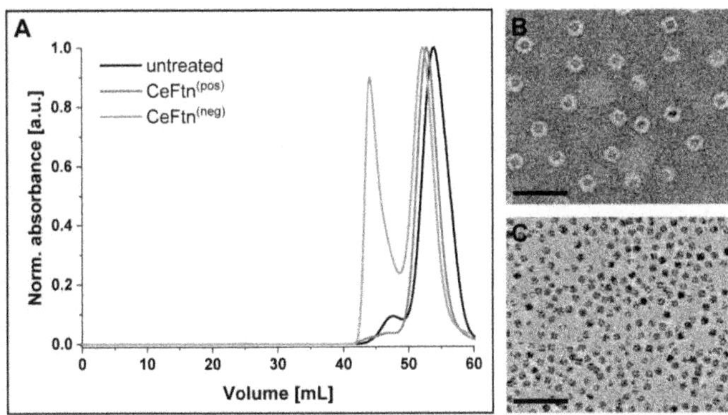

Figure 5.41: Synthesis of CeFtn$^{(pos)}$ and CeFtn$^{(neg)}$. (A) Size exclusion chromatogram of untreated protein container (black), CeFtn$^{(pos)}$ (blue) and CeFtn$^{(neg)}$ (red). TEM images of CeFtn$^{(pos)}$ stained (B) and unstained (C). Scale bars are 50 nm.

Moreover, the synthesis of iron oxide nanoparticles was optimized. In previous work d-spacing of the iron oxide nanoparticles showed a mixed composition.[189] Presumably, the nanoparticles contained different phases. The aim was the synthesis of pure ferrimagnetic magnetite (Fe_3O_4) iron oxide nanoparticles. Schwarzacher *et al.* successfully synthesized magnetite nanoparticles in horse spleen apoferritin.[197,264] Moreover, the crystallization of the sample was feasible to obtain nanoparticle filled protein crystals showing magnetic properties. This synthesis was adapted for Ftn$^{(neg)}$. Compared with the previous synthesis the pH value was increased to 8.6 and the flow rate decreased to 9 ions per minute per ferritin cage. The sample was purified by several centrifugation steps to separate filled containers from free iron oxide nanoparticles. These nanoparticles formed a pellet during the centrifugation. The pellet was resuspended in water and movement of the particles in vicinity of a strong neodymium magnet was observed. This observation indicated the formation of Fe_3O_4 nanoparticles outside the protein container. The protein container sample was further purified by SEC (Figure 5.42A). Presence of iron oxide was monitored at an absorbance of 322 nm. Two discrete peaks were obtained at elution volumes of 45 mL and 54 mL. The second peak eluted at the same volume as $Ftn^{(neg)}$ and contained nicely formed protein containers. The high absorbance at 322 nm indicated a loading with iron oxide nanoparticles. The first peak at 45 mL belonged to larger aggregates because of co-precipitation of the protein with nanoparticles while iron oxide nanoparticles were formed in the solution. Empty $Ftn^{(neg)}$ was separated from $FeFtn^{(neg)}$ by sucrose gradient centrifugation (Appendix, Figure 8.10). After the purification, the solution showed a characteristic brown color indicating the incorporation of iron oxide nanoparticles inside $Ftn^{(neg)}$. Moreover, ferrofluidic behavior was observed in the vicinity of a strong neodymium magnet. TEM images (Figure 5.42B and C) showed the presence of ferritin container and iron oxide nanoparticles.

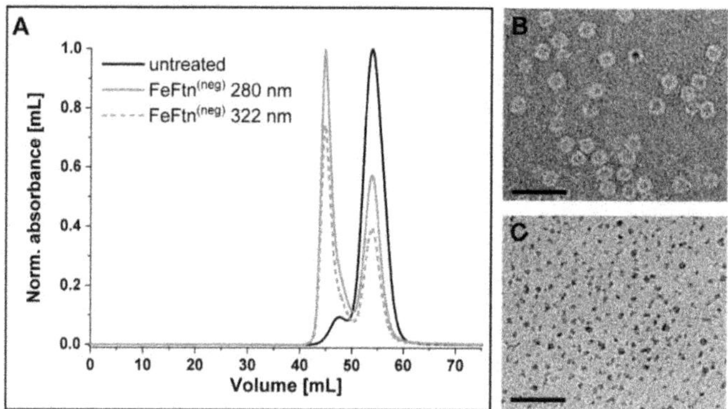

Figure 5.42: Synthesis of $FeFtn^{(neg)}$. (A) Size exclusion chromatogram of untreated protein container (black) and $FeFtn^{(neg)}$ (red). TEM images of $FeFtn^{(neg)}$ stained (B) and (C) unstained. Scale bars are 50 nm.

All in all, cerium oxide nanoparticles were synthesized *in situ* in both protein containers. Moreover, the synthesis of iron oxide nanoparticles inside of $Ftn^{(neg)}$ was optimized to obtain magnetic Fe_3O_4 nanoparticles.

5.4 Assembly of the building blocks

For the assembly of the nanoparticle-protein container composites, it is crucial to carefully control the attractive and repulsive forces. On the one hand, strong or irreversible interactions of the containers will not lead to crystal formation and favor gel-like aggregates. On the other hand, weak interactions will lead to no assembly at all. The assembly of the oppositely charged protein containers is mediated by attractive interactions between the complementary charged containers, such as electrostatic interactions and hydrogen bonds between the side chains of the amino acids. These interactions are mainly influenced by the ionic strength of the solution (Figure 5.43).

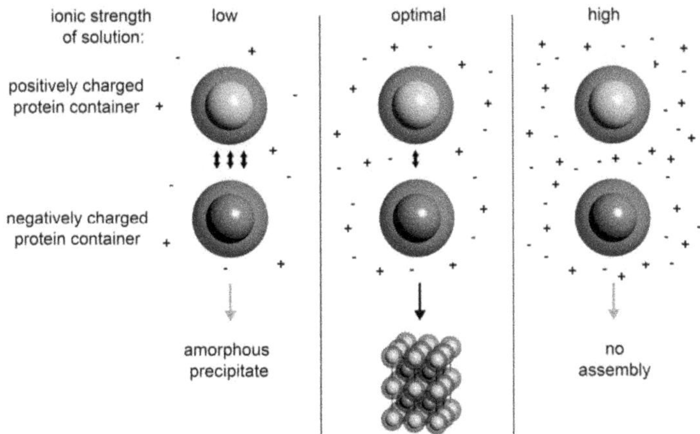

Figure 5.43: Assembly of the building blocks. Influence of the ionic strength of the media on binary crystal assembly. The interaction between the charged protein containers can be controlled with ionic strength (salt concentration) in the crystallization conditions. Figure adopted from Tobias Beck.

Additionally, the pH value determines the ionization state of the amino acid residues and thus the total surface charge of the protein containers. In previous works, four different crystal structures were established.[40,189,201] Two binary structures and one unitary structure for each protein container were determined. Moreover, the assembly of metal oxide filled protein container was already investigated. Hence, these results will be used as starting point.

This chapter is divided into two parts. In the first part, the crystallization of empty protein containers will be further investigated. Batch crystallization was performed to increase the yield of crystals. The second part will be concerned with the assembly of the AuNP-filled protein containers.

5.4.1 Batch crystallization

To crystallize the protein container, usually hanging drop vapor diffusion was used as crystallization method (Figure 5.44A). In detail, a drop combining protein solution and reservoir solution is prepared and placed above the reservoir solution in a closed system. The precipitant concentration difference in the drop and reservoir solution leads to diffusion of water until both solutions, drop and reservoir, have reached an equilibrium. During this process crystals will grow (see below). To obtain a high amount of crystals, the procedure must be repeated several times due to the limitation in drop size because the drop must hang above the reservoir solution. In contrast, the only limitation for batch crystallization is the size of the crystallization chamber. Here, protein solution is directly mixed with the whole reservoir solution (Figure 5.44B). With respect to the phase diagram both methods show different crystallization pathways (Figure 5.44C).

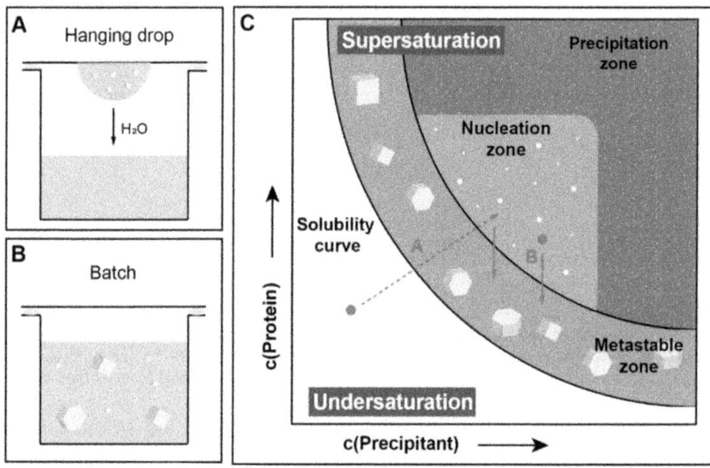

Figure 5.44: Crystallization methods. (A) Schematic illustration of the hanging drop crystallization method. (B) Schematic illustration of the batch crystallization method. (C) Phase diagram for the crystallization of macromolecules. The crystallization route is marked in red for each method.[182]

In the hanging drop method, the drop starts in the stable undersaturated phase, where no interactions between protein containers occur. Due to diffusion of water the precipitant concentration increases inside the drop to reach the nucleation zone. The formation of nuclei and growth of the nuclei happen simultaneously. After nuclei are formed, the metastable zone will be reached where crystal growth occurs, but no new nuclei are formed. On the other hand, in the batch crystallization method, the starting point is located directly in the nucleation zone. Again, after nuclei are formed the metastable zone will be reached for crystal growth.[182,265–267]

Batch crystallization experiments were conducted based on a method by Rayment.[268] In detail, protein solution was transferred to an *Eppendorf* tube and vortexed gently while the crystallization condition was added dropwise. Subsequently, the solution was transferred to a greased batch plate and seal with a cover slide. For the crystallization, one unitary structure (PDB:5JKK) and one binary structure (PDB:5JKL) were chosen for two reasons. First, the optimization of the procedure is easier for the unitary structure because of less components during the crystallization experiment. Afterwards, it will be transferred to obtain binary crystals. Second, both crystallization conditions are very similar. By changing the magnesium ion concentration both crystal structure can be obtained.[201] Experiments were performed with 5 µL protein solution and a total volume of 20 µL. Compared with the hanging drop method, a 5-fold excess was used. First experiments with the same protein concentration (4 mg/mL) and precipitant concentration (500 mM Mg^{2+}) as the hanging drop method led to small crystals (Appendix, Figure 8.11). Due to the different crystallization pathway, an additional optimization of the crystallization condition must be performed. The protein concentration was increased to 10 mg/mL and the Mg^{2+} concentration was screened between 100 mM and 600 mM.

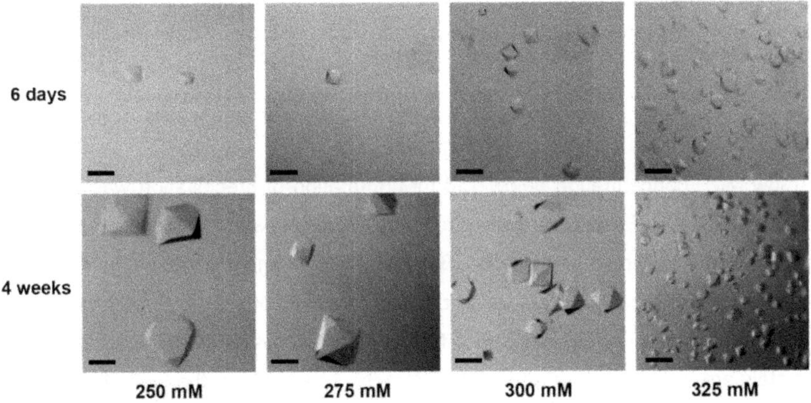

Figure 5.45: Screening of Mg^{2+} concentration for unitary structure. Ftn(neg) crystals grown in microbatches with Mg^{2+} concentration between 250 mM and 325 mM after 6 days and 4 weeks crystallization time. Scale bars are 200 µm.

No crystal growth was observed between 100 mM and 250 mM. The largest crystals with a size up to 400 µm were obtained at Mg^{2+} concentration between 250 mM and 275 mM (Figure 5.45). A further increase of Mg^{2+} concentration led to smaller crystals. Even at 600 mM Mg^{2+} concentration crystal growth was observed. Comparison of crystallization conditions from both methods showed a twofold higher Mg^{2+} concentration was needed for the hanging drop method. Crystallization studies of both methods with several other proteins showed a similar correlation.[267] In microbatch, the concentration of precipitant is 10-20% lower than that in the reservoir of hanging drop experiment. Here, the precipitant concentration is even 50% lower. Moreover, the required time for crystal growth differed strongly. For the hanging drop method, crystals appeared after 2 days and did not change their size significantly by increasing the time. In contrast, for batch crystallization smaller crystals appeared after 6 days, which grew significantly larger after 4 weeks.

The same procedure was adapted for the crystallization of binary crystals. Here, the protein concentration was directly adjusted to 10 mg/mL for both protein variants. Mg^{2+} concentration was screened between 100 mM and 350 mM.

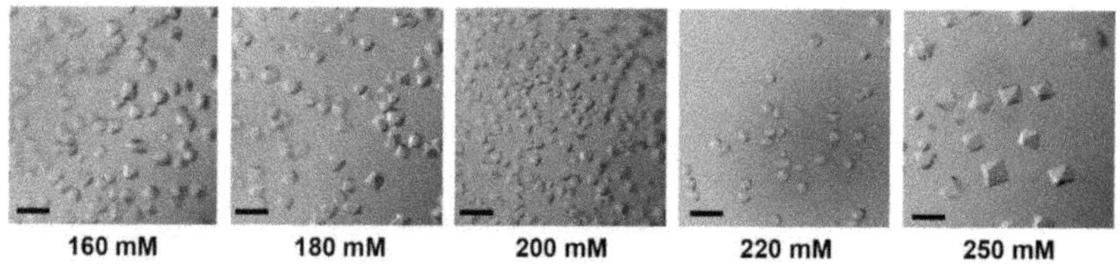

Figure 5.46: Screening of Mg^{2+} concentration for binary structure. Crystals grown in microbatches with Mg^{2+} concentration between 160 mM and 250 mM using both protein variants. Scale bars are 200 µm.

At low Mg^{2+} concentration below 130 mM, formation of precipitation was observed. Due to the low Mg^{2+} concentration, strong interaction between both protein containers occurred resulting in co-precipitation. Between 130 mM and 350 mM octahedral crystal appeared. The largest crystals were obtained between 250 mM and 350 mM. Unfortunately, the unitary crystal structure is formed at the same condition. Presumably, the crystals are only composed of $Ftn^{(neg)}$. From the screening for the unitary structure, it is known that unitary crystals are only form above a Mg^{2+} concentration of 250 mM. Below a Mg^{2+} concentration of 250 mM, no crystal formation (clear drop) occur. Thus, in batch crystallization, at 180 mM binary crystals with a size of 150 µm were obtained (Figure 5.46). A similar Mg^{2+} concentration was used as for the hanging drop crystallization (200 mM). Moreover, no difference was observed in crystal size by increasing the crystallization time from 1 day to 4 weeks (Appendix, Figure 8.12). In batch crystallization, the unitary Ftn$^{(neg)}$ crystals grew larger over time until reaching a final size of 400 µm. Similar to the hanging drop crystallization, growth was finished after one day for the binary structure in the batch crystallization.

Finally, a last increase of crystallization volume was performed from 20 µL to 50 µL to investigate if the batch crystallization is independent from the crystallization volume. Compared with the hanging drop method, this equals twelve drops or half of a crystallization plate, respectively. Both protein containers were crystallized with a concentration of 10 mg/mL in 180 mM Mg^{2+} concentration. Again, well-shaped octahedral crystals were obtained (Figure 5.47).

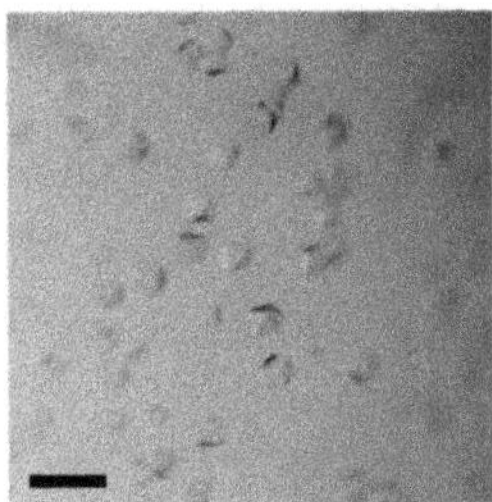

Figure 5.47: Upscaling of the batch crystallization. Crystallization of binary crystals with a total volume of 50 µL. Scale bar is 200 µm.

As the upscaling did not have any impact on the crystallization, it is possible that the crystallization can be upscaled even further without loss in crystal quality. The high order of the protein containers in the protein crystal was confirmed by SAXS measurements of a crystal ensemble (Appendix, Figure 8.13 and Table 8.2). This fact will increase the efficiency of the material production.

In summary, batch crystallization was performed to obtain unitary crystals as well as binary crystals. Moreover, upscaling was possible to 50 µL crystallization volume without loss of crystal quality.

5.4.2 Formation of nanoparticle superlattices

As the final step towards construction of highly ordered materials, the obtained nanoparticle-loaded building blocks were crystallized using the crystallization conditions for empty protein containers. In previous works with metal oxide loaded protein containers, the nanoparticle cargo did not affect the crystallization of the protein containers. The same crystal structures were obtained as for empty protein containers.[40] For crystallization 200 mM Mg formate as crystallization condition was chosen to obtain the binary structure. In total, three different types of cargo were incorporated in the protein containers: plasmonic nanoparticles (5.1), dye molecules (5.2) and metal oxide nanoparticles (5.3). First, the crystallization of each protein container composite was investigated separately. Ideally, octahedral crystals with a size larger than 100 µm will be obtained. In the final step, different types of cargo were combined to obtain binary materials.

The encapsulation of plasmonic nanoparticles was performed with two different conditions. The crystallization of empty protein containers after reassembly revealed different crystallization behavior (5.1.3). Protein containers reassembled with the Gua condition were able to crystallize without problems. Even large octahedral crystals were obtained directly after reassembly without any further purification by SEC. Crystallization of protein containers reassembled with pH 2 showed strong dendritization. A second run with SEC was mandatory to obtain nicely shaped octahedral crystals. First, the crystallization of samples treated with the Gua condition was studied (Figure 5.48).

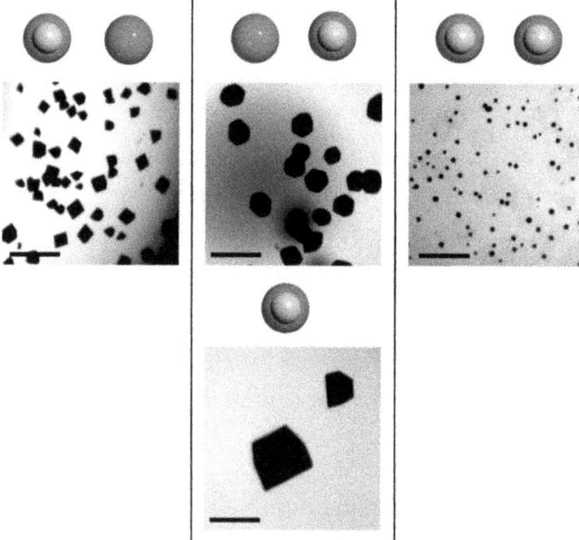

Figure 5.48: Optical microscopy images of plasmonic nanoparticle filled protein crystals. Crystallization of different combinations of plasmonic nanoparticle filled protein containers. Scale bars are 200 µm.

Each filled protein container was crystallized with the oppositely charged empty protein container (eFtn$^{(neg)}$, eFtn$^{(pos)}$) before the crystallization was performed with both AuNP filled protein containers. Moreover, AuFtn$^{(neg)}$ was crystallized with the crystallization condition for a unitary structure. All combinations yielded crystals, but the size of the crystal varied for all samples. The crystal size decreased in the order eFtn$^{(pos)}$/AuFtn$^{(neg)}$ > AuFtn$^{(pos)}$/eFtn$^{(neg)}$ > AuFtn$^{(pos)}$/AuFtn$^{(neg)}$. The decrease in crystal size indicated an influence of encapsulation for the crystal growth. If only one AuNP filled component was used, the crystal growth was not affected due to the untreated protein containers. Small alterations of protein container structure in the AuNP filled variants were tolerated by the untreated protein container. By combining two AuNP filled components, every alteration of the protein container can affect the crystal growth. The AuFtn$^{(pos)}$/AuFtn$^{(neg)}$ only grew to a size of approximately 30 µm. Moreover, crystallization of unitary AuFtn$^{(neg)}$ crystals was possible. Theoretically, the same hampered crystal growth as for AuFtn$^{(pos)}$/AuFtn$^{(neg)}$ crystals would be expected. However, the largest crystals with a crystal size up to 200 µm were obtained. This observation can be explained by the difference in crystal contact interfaces in both crystal structures. In the binary structure Ftn$^{(pos)}$ and Ftn$^{(neg)}$ interact with each other by hydrogen bonds or salt bridges of certain amino acids on the outer surface. Small alterations can weaken the interaction and terminate further crystal growth. In the unitary structure two Ftn$^{(neg)}$ are connected through complexation with a magnesium ion. Probably, the crystal interfaces in the unitary structure are less affected by alteration of the protein container during the encapsulation. One reason could be the stronger bond between the amino acids and the metal ions in a coordination complex compared to salt bridges and hydrogen bonds in the binary structure. The encapsulation with the pH 2 condition was only possible for Ftn$^{(neg)}$. Binary crystallization experiments with samples after one SEC purification led to strong dendrimer-like structures (Appendix, Figure 8.14). However, further purifications did not improve the crystal growth to obtain nicely shaped octahedral crystals. Usually, the crystal had at early stages of crystal growth an octahedral morphology, but with increasing time one lattice direction was favored. Thus, elongated rod-like crystal with the crystal habit of an elongated dodecahedron grew. In contrast, the crystallization of the unitary structures led again to large octahedral crystals up to 400 µm in size. Similar to the Gua condition, the crystal grow of the unitary Ftn$^{(neg)}$ structure is less affected by alteration of the protein container during the encapsulation.

Protein crystals are fragile and shatter easily under mechanical pressure. The generation of stable crystals is crucial for further structural characterization and potential applications of the material. Regular protein crystals can be readily stabilized by cross-linking with a suitable agent.[269,270] For example, dialdehydes have been frequently used for crystal fixation to reduce the impact of cryo-cooling required for synchrotron data collection.[271] For binary ferritin crystals, this procedure was applied to enhance the stability of the crystals. The protein shells were cross-linked with glutaraldehyde after crystal formation by simply adding the cross-linking agent to the reservoir solution. The aldehyde groups of the crosslinker reacts with lysine residues of the protein container to form an imine. After the cross-linking, an easy manipulation of the crystals is feasible without destroying the crystals. Residual glutaraldehyde is washed away with pure water and the cross-linked crystals can be dried to obtain free-standing crystals. In this way, crystals are yielded, which can be used for investigations with an electron microscope. AuFtn$^{(pos)}$/eFtn$^{(neg)}$ crystals were cross-linked and investigated by SEM analysis (Figure 5.49).

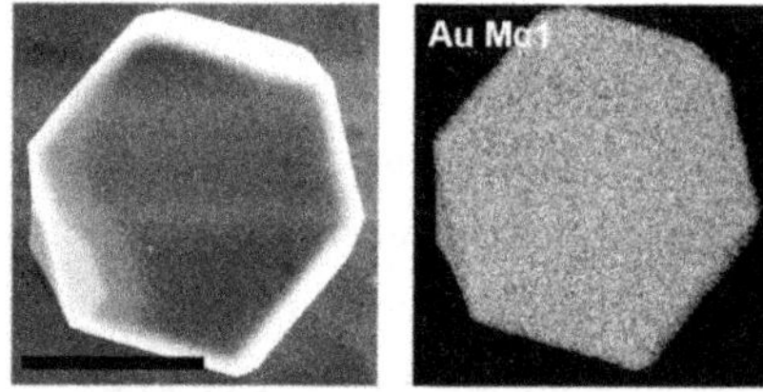

Figure 5.49: SEM measurement of AuFtn$^{(pos)}$/eFtn$^{(neg)}$ crystal and EDX mapping. SEM image of AuFtn$^{(pos)}$/eFtn$^{(neg)}$ crystal (left) and EDX mapping of the crystal (right). Scale bar is 25 µm.

EDX mapping of the crystal showed the homogenous incorporation of AuNP in the crystal. Moreover, AuFtn$^{(pos)}$/eFtn$^{(neg)}$ crystals were further characterized by SAXS measurements to determine the nanoparticle superlattice structure (Figure 5.50). For the SAXS analysis, stabilized crystals from one crystallization plate were washed in water, centrifuged and collected in a capillary. The capillary was rotated in the X-ray beam to ensure the diffraction of the ensemble and not of only few crystals, similar to X-ray powder diffraction. Centric diffraction rings were obtained due to different orientation of the crystals in the ensemble. Moreover, a high long-range order and large domain size of the crystals were observed by strong distinct Bragg peaks within the diffraction rings. From the 2D SAXS pattern 1D SAXS data were obtained. The nanoparticle formed a primitive tetragonal lattice (P4/mmm, space group no: 123) composed of one single phase. From the peak position of the first two reflections (001) and (100), unit cell parameters were calculated (see section 7.5.11). The obtained cell parameters were used to simulate the respective full diffraction pattern (red line, Figure 5.50).

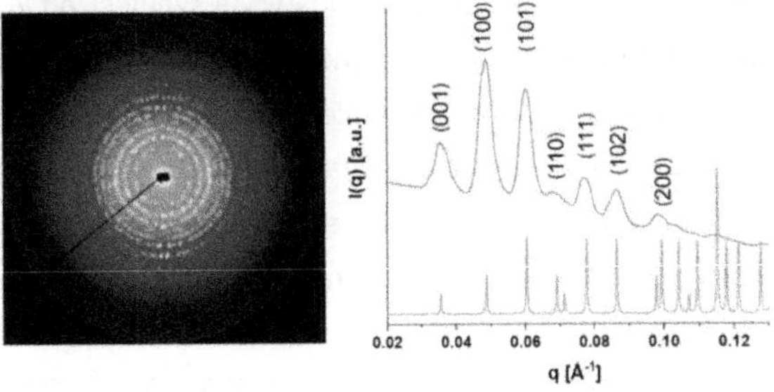

Figure 5.50: SAXS data of AuFtn$^{(pos)}$/eFtn$^{(neg)}$. 2D SAXS pattern (left) and radially averaged 1D SAXS data (right) of AuFtn$^{(pos)}$/eFtn$^{(neg)}$. The measured data are displayed in black and simulated data in red.

In Table 5.10 the measured and simulated peak positions are compared. The calculated values for the diffraction fit very well with the measured one. The unit cell parameters were approximately 1% larger than for the X-ray crystal structure (a = 126.6 Å, c = 174.9 Å). The shrinking of the cell for the X-ray crystal structure can be explained by different temperatures during both measurements: SAXS data collection was performed at room temperature and the X-ray data collection at 100 K.

Table 5.10: Cell parameters obtained from SAXS data.

hkl	experimental q (Å^{-1})	calculated q (Å^{-1})	cell parameters
001	0.03561		c = 176.4 Å
100	0.04883		a = 128.7 Å
101	0.06010	0.06044	
111	0.07721	0.07770	
102	0.08655	0.08635	

In a next step, dye-loaded protein containers were crystallized (Figure 5.51I-II). Successful encapsulation was visible due to the characteristic color of the protein crystals depending on the dye. Crystallization of metal oxide nanoparticle-loaded protein containers was proven in previous work. Here, the crystallization of the new iron oxide loaded protein container was shown (Figure 5.51III-V). Combination with CeFtn$^{(pos)}$ is also possible to obtain a binary system composed of two different nanoparticles. Finally, AuNP filled protein containers were combined with dye filled protein containers as well as metal oxide nanoparticle filled protein containers to obtain new binary materials (Figure 5.51VI-IX). All combinations yielded nicely shaped crystal with a polyhedron morphology. Importantly, combinations of different building blocks independent of the encapsulation strategy are feasible. AuNPs incorporated with the encapsulation strategy were combined with metal oxide nanoparticles synthesized *in situ* in the protein container cavity. Crystals composed of one filled component possessed an octahedral crystal habit similarly to the empty protein crystals. However, combinations of two filled protein containers affected the crystal quality. Crystal growth along specific lattice direction was more favored resulting in elongated crystals facet in the shape of a kite. Nevertheless, the sharp edges and distinct form of the protein crystals indicated a high order in the material.

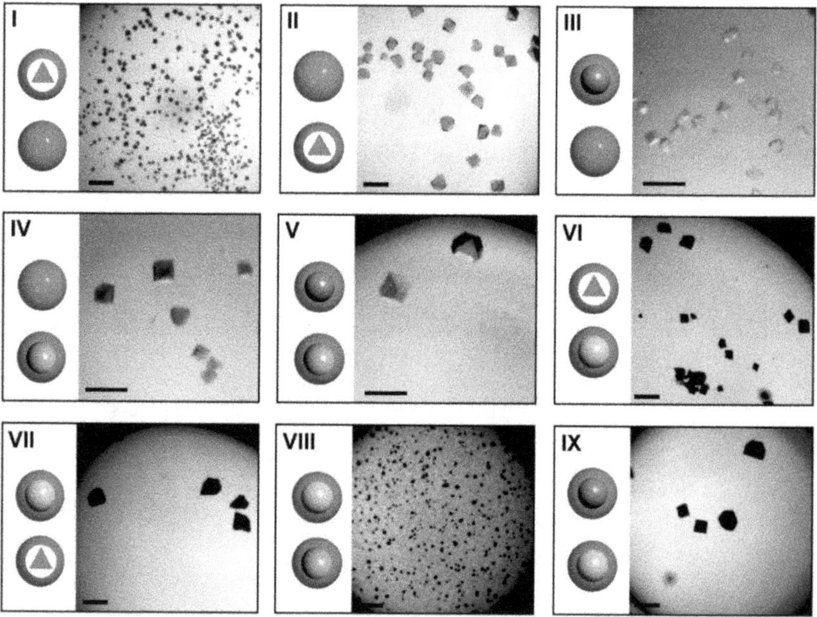

Figure 5.51: Optical microscopy image of protein crystal with different cargo. (I-II) Dye molecule filled protein containers, (III-V) metal oxide nanoparticle filled protein containers, (VI-VII) combination of dye and AuNP filled protein containers and (VIII-IX) combination of metal oxide nanoparticle and AuNP filled protein containers. Scale bars are 200 µm.

As a result, the crystallization of AuNP-filled protein containers to obtain the highly ordered material was successful. SAXS analysis of the material confirmed the high order of the material. Moreover, the combination with different types of cargo was performed to create new nanomaterials.

5.5 Potential applications of the nanomaterial

The precise arrangement of nanoparticles in defined structures is a prerequisite for the generation of so-called metamaterials. Optical properties of the plasmonic nanoparticles were investigated in the nanoparticle superlattice to determine the influence of the interaction between nanoparticles for the properties. Moreover, catalytic activity of encapsulated metal oxide nanoparticles as well as AuNPs was studied. The catalytic performance in solution was compared to the catalytic reactions carried out in the protein crystals.

5.5.1 Optical properties

Plasmonic materials based on protein containers could find future application in sensing: The well-defined channels in the crystal lattice, which can be tuned with the protein scaffold, and the charges on the container surface can act as filters for analytes such as small molecules. Binding of molecules close to the AuNP on the outer container surface could possibly be detected by monitoring the plasmon absorption spectra. Moreover, localized surface plasmon resonances induce a significant enhancement of the electromagnetic field close to the surface of the nanoparticles.[85] As the resonant frequency depends on the refractive index of the environment, frequency shifts are used as a probe to determine the concentration of the molecules.[95,272–274] Moreover, these nanomaterials can be used as substrate in SERS to enhance Raman scattering of molecules.[97,275] To observe these properties, short distances between AuNPs must be present to ensure strong interactions between two different building blocks. To quantify the strength of the interaction between building blocks mixed $AuFtn^{(pos)}/RhFtn^{(neg)}$ crystals were used. Plasmon-exciton coupling is observed if plasmonic nanoparticles interact with fluorophores. The type of coupling is strongly dependent on the strength of the interactions or distance between both components, respectively. In a first step, the composition of the crystal was investigated. Emission spectra of $eFtn^{(pos)}/eFtn^{(neg)}$, $eFtn^{(pos)}/RhFtn^{(neg)}$, $AuFtn^{(pos)}/eFtn^{(neg)}$ and $AuFtn^{(pos)}/RhFtn^{(neg)}$ crystals were measured at an excitation wavelength of 405 nm (Figure 5.52A). Comparison of the spectra proved the binary composition of $AuFtn^{(pos)}/RhFtn^{(neg)}$. Emission maxima of both components were present at the same position as for the unitary loaded crystals. In a second step, imaging of the crystals was carried out with a confocal microscope to visualize the fluorescence at an excitation wavelength of 405 nm (Figure 5.52B and D). As expected, the $eFtn^{(pos)}/RhFtn^{(neg)}$

crystal showed a strong fluorescence signal. This crystal was used to calibrate the microscope for further measurements. Confocal microscope images of AuFtn$^{(pos)}$/eFtn$^{(neg)}$ and AuFtn$^{(pos)}$/RhFtn$^{(neg)}$ crystals showed no fluorescence. AuNPs possess no fluorescence at an excitation wavelength of 405 nm. Moreover, a quenching of the rhodamine fluorescence in the binary crystal was observed. As control experiment, empty protein crystals were measured to ensure the applicability of the setup. Weak fluorescence was observed, caused by the amino acids tryptophan, tyrosine, and phenylalanine in the protein container. Consequently, the quenching was caused by the AuNPs. The quenching of the dye can be explained by an energy transfer (FRET).[276–278] The dye rhodamine B functions as donor and transfer its energy non-radiatively to the AuNP (acceptor), resulting in a decrease of the fluorescence (plasmon-exciton coupling). Typically, FRET is observed for dye-dye interaction and is limited to distances up to 100 Å. Moreover, the rate of the energy transfer is strongly dependent on the donor-acceptor distance because it is based on dipole-dipole interactions (inverse sixth power distance dependency). Thus, the distances in the protein crystals are adequate to observe a coupling between the AuNPs and dye molecules. However, further measurements are required to investigate the interactions in the protein crystal more in detail.

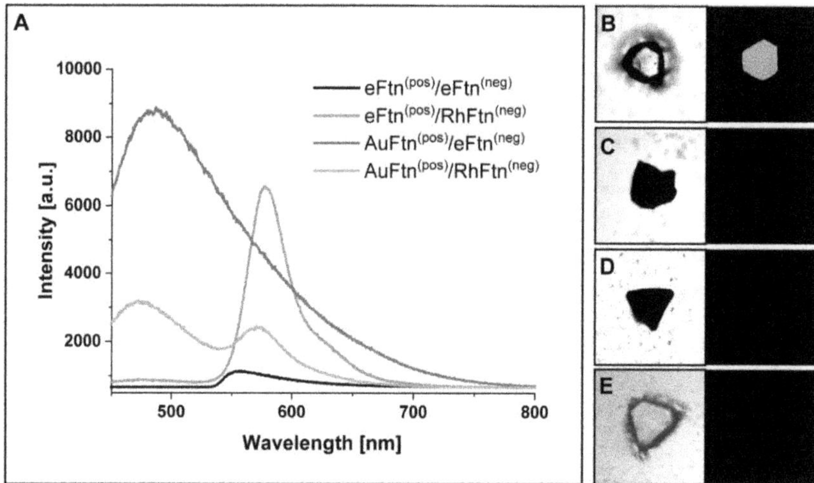

Figure 5.52: Plasmin-exciton coupling in protein crystals. (A) Emission spectra at an excitation wavelength of 405 nm of eFtn$^{(pos)}$/eFtn$^{(neg)}$ (black), eFtn$^{(pos)}$/RhFtn$^{(neg)}$ (red), AuFtn$^{(pos)}$/eFtn$^{(neg)}$ (blue) and AuFtn$^{(pos)}$/RhFtn$^{(neg)}$ (green). Confocal microscopy and corresponding fluorescence (λ = 405 nm) images of (B) eFtn$^{(pos)}$/RhFtn$^{(neg)}$, (C) AuFtn$^{(pos)}$/eFtn$^{(neg)}$, (D) AuFtn$^{(pos)}$/RhFtn$^{(neg)}$ and (E) eFtn$^{(pos)}$/eFtn$^{(neg)}$.

Subsequently, the application of the protein crystals as a SERS substrate was investigated. 4-Aminothiophenol (4-ATP) was chosen as molecule for the detection. Raman investigations were carried out with a concentration of 1.0 M 4-ATP dropped on three different substrates at an excitation wavelength of 633 nm: glass, a AuNP solution dropped on glass, and AuNP-loaded crystal on glass. The same settings were used to compare the effect of the substrate on the Raman measurement.

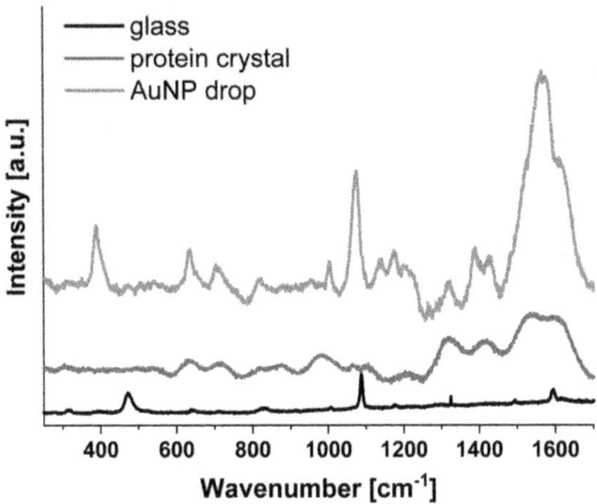

Figure 5.53: SERS measurements of 4-ATP with different substrates. 4-ATP was measured on glass (black), AuNP-loaded protein crystals (blue) and AuNP drop (red).

Figure 5.53 shows the Raman spectra for the three substrates. In the measurement on glass, three strong Raman bands at 473, 1087 and 1590 cm^{-1} exist, which corresponds to the vibration modes of $\gamma_{CCC}(a_1)$, $\nu_{CS}(a_1)$ and $\nu_{CC}(a_1)$.[279,280] Raman bands at 828, 1006 and 1322 cm^{-1} were weakly visible. Measuring a crystal with 4-ATP showed no sharp distinct signals. Seven very weak broad signals were distinguished. However, only the vibration mode of $\nu_{CC}(a_1)$ was present, shifted to 1602 cm^{-1}. In contrast, the Raman spectrum measured on the AuNP drop showed sharp signals with the highest intensities. Raman bands at 819 and 1571 cm^{-1} were referred to vibration modes of $\nu_{CS}(a_1)$ and $\nu_{CC}(b_1)$. Interestingly, Raman bands at 634 and 703 cm^{-1} occurred only in the spectra measured with AuNP drop and crystal as substrate (Table 5.11). Moreover, weak signals at 1106 and 1322 cm^{-1} were enhanced with both AuNP-based substrates.

Table 5.11: Comparison of the position of Raman bands in different measurements.

Literature[a,279]	**Glass**[a]	**Protein crystal**[a]	**AuNP drop**[a]	**Assignment**[abc]
396			389	γ_{CC} (a_2)
463	473			γ_{CCC} (a_1)
634		631	634	γCCC (a_1)
699		712	703	$\pi_{CH}+\pi_{CS}+\pi_{CC}$ (b_1)
820	828		819	π_{CH} (b_1)
1011	1006	981	1003	$\gamma_{CC}+\gamma_{CCC}$ (a_1)
1089	1087		1076	ν_{CS} (a_1)
1142			1138	δ_{CH} (b_2)
1173			1176	δ_{CH} (a_1)
1310	1322	1319	1319	$\nu_{CC}+\delta_{CH}$ (b_2)
1403		1416	1390	
1445			1429	$\nu_{CC}+\delta_{CH}$ (b_2)
1572			1571	ν_{CC} (b_2)
1595	1590	1602		ν_{CC} (b_1)
1620			1613	δ_{NH}

[a] Frequencies are quoted in cm^{-1} [b] Approximate description of the modes (ν = stretch, δ and γ = bend, π = wagging) [c] Benzene ring vibrations are classified as a_1, a_2, b_1 and b_2

The comparison of the spectra led to two consequences. First, the synthesized AuNPs can enhanced the Raman signal of molecules as SERS substrate. Nevertheless, for AuNPs only dropped on the glass substrate without ensuring a certain order, an enhancement factor of 1.5 was calculated by comparing the intensities of the ν_{CS}(a_1) vibrational mode. Second, a Raman spectrum with broad and weak signals was collected on the protein crystal as substrate. Data collection on protein crystals without substrate led to no signals (Appendix, Figure 8.15). However, it seemed that the organization of AuNP-filled protein containers into a 3D crystal is not beneficial for the use as a SERS substrate. The enhancement of Raman signals depends on the size[281] and shape[282] of the nanoparticle as well as the thickness[283] and roughness[284] of the material. Investigations on AgNPs showed an increase of enhancement factor by increasing the roughness of the sample.[284] Moreover, an optimal thickness exists, where at higher values the enhancement factor decreases again. Presumably, a coating with thin layers of the AuNP-loaded protein container building blocks would lead to superior materials for application as SERS substrate.

5.5.2 Catalytic activity

Another interesting field of application for the nanostructured material is catalysis. The fine control over material composition and the defined structure is advantageous for reaction control. Importantly, the high solvent content in the protein crystals indicates a high porosity of the material enabling the access to the reactive nanoparticles through pores inherent to the protein containers.[285] Due to the binary character of the material, two different types of reactions can be performed simultaneously. Cerium oxide nanoparticles are known for their enzyme-like activity.[204,286] Nanoparticles can be superior in terms of thermal stability and resistance against degradation. However, reactivity of cerium oxide nanoparticles is highly dependent on the composition between Ce^{3+} and Ce^{4+}.[287] Powder X-ray diffraction of CeFtn$^{(pos)}$ showed that the particles inside the protein containers possess the same crystal structure as commercially available CeO_2 (Appendix, Figure 8.16). Here, the oxidase and peroxidase-like activity was investigated by the catalytic oxidation of 3,3',5,5'-tetramethylbenzidine (TMB), a substrate for horse-radish peroxidase commonly used in bioassays (Appendix Figure 8.17).[288] The oxidation of TMB can be easily detected because of the traceable blue oxidation product: Since cerium oxide nanoparticles do not show any strong absorbance in the visible range, the formation of the product can be followed by monitoring the absorbance maximum at 645 nm with UV-Vis spectroscopy. Before the protein crystal was used as heterogenous catalyst, the catalytic reaction was studied in solution. In detail, reaction was started by mixing the substrate TMB with either CeFtn$^{(pos)}$ or CeFtn$^{(neg)}$. Oxidase-like activity investigations were conducted under aerobic conditions. To study the peroxidase-like activity, hydrogen peroxide was added to the reaction mixture. Additionally, control experiments without catalysts were carried out. Figure 5.54A and B show the UV-Vis spectra for the oxidase-like and peroxidase-like activity, respectively. In general, the blue colored product was obtained for CeFtn$^{(pos)}$ and CeFtn$^{(neg)}$ with and without hydrogen peroxide. Without catalyst no background reaction was observed. However, the reaction proceeds at a faster rate in the presence of hydrogen peroxide. Here, the blue color appeared within minutes. On the contrary, an incubation time of one day was required for the reaction under aerobic condition, with oxygen as oxidation agent.

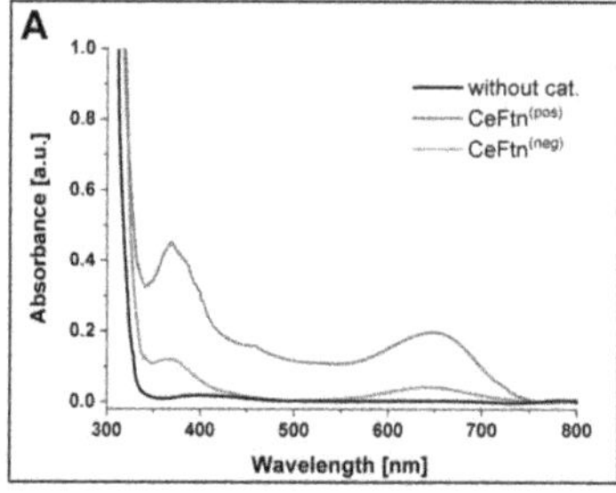

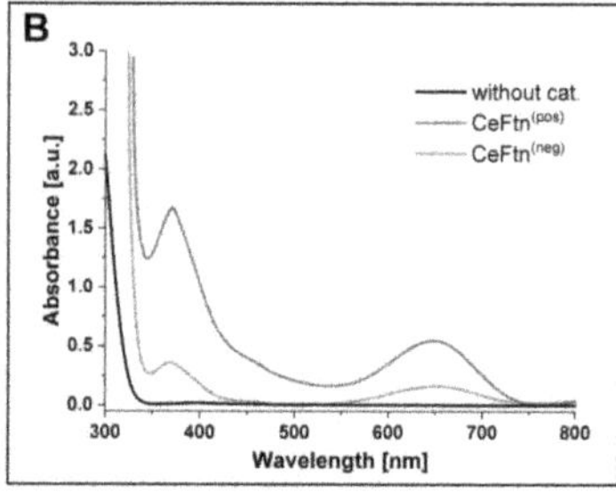

Figure 5.54: Catalytic activity of CeFtn$^{(pos)}$ and CeFtn$^{(neg)}$ in solution. Absorption spectra of TMB after catalytic reaction with CeO_2 in Ftn$^{(pos)}$ (blue) and Ftn$^{(neg)}$ (red): (A) oxidase-like activity after one day and (B) peroxidase-like activity in the presence of hydrogen peroxide after an incubation time of 15 min. Control experiments without catalyst in black.

Kinetic parameters for the oxidase-like and peroxidase-like activity of both protein containers were determined (Table 5.12, Appendix Figure 8.18). For oxidase-like activity, the concentration of TMB was varied while the concentration of nanoparticle-loaded protein container was kept constant. Kinetic parameters for the peroxidase-like activity were measured at constant concentrations of TMB and nanoparticle-loaded protein containers while the concentration of hydrogen peroxide was varied.

Table 5.12: Comparison of kinetic parameters for the oxidase-like and peroxidase-like activity.

Building block	Substrate	v_{max} [µM s^{-1}]	K_m [mM]
Ftn$^{(pos)}$	TMB	2.2×10^{-3}	1.54
Ftn$^{(pos)}$	TMB + H_2O_2	1.2×10^{-2}	15.39
Ftn$^{(neg)}$	TMB	7.3×10^{-4}	0.83
Ftn$^{(neg)}$	TMB + H_2O_2	3.3×10^{-3}	4.05

Comparison of the kinetic parameters showed a higher activity of CeFtn$^{(pos)}$ compared to CeFtn$^{(neg)}$ and a higher reaction velocity in the presence of hydrogen peroxide (peroxidase-like activity) for both protein containers. The difference in activity can be attributed to a higher nanoparticle loading of CeFtn$^{(pos)}$: Analysis of the sucrose gradient centrifugation showed a shift of the peak for the maximum absorption at 322 nm to higher fractions for CeFtn$^{(pos)}$ compared to CeFtn$^{(neg)}$, indicating a higher mass of CeO_2 nanoparticles inside the protein container cavity. Additionally, the higher loading efficiency was confirmed by higher absorption of CeFtn$^{(pos)}$ during UV-Vis measurements of both protein containers at the same protein concentration (Appendix, Figure 8.19). The higher reaction velocity in presence of hydrogen peroxide is presumably caused due to different reaction pathways. On the one hand, under aerobic condition (oxidase-like activity), the substrate needs to diffuse into the protein container through the pores to be oxidized on the particle's surface. On the other hand, in the presence of hydrogen peroxide (peroxidase-like activity), the substrate is probably oxidized by hydroxyl and superoxide radicals formed by conversion of hydrogen peroxide at the nanoparticle surface in a Fenton-like reaction.[289] For ferritin loaded with platinum nanoparticles a similar trend in reaction velocity and K_m was observed for the reaction with and without hydrogen peroxide.[290] Interestingly, comparing the kinetic parameters for CeFtn composites with cerium oxide nanoparticles coated with polymer ligands[291] similar Michaelis constants were obtained, indicating a similar binding affinity to the nanoparticle surface for both types of catalysts. However, the reaction velocity is at a factor 10^3 slower for the protein-nanoparticle composites because of the additional diffusion of the substrates through the pores before and after the reaction.

After the successful characterization in solution, the reaction was performed in the nanoparticle-loaded protein crystals to observe if the catalytic activity is preserved in the nanoparticle superlattice. Prior to the investigations, protein crystals were cross-linked with glutaraldehyde to enhance the stability of the protein crystals. Incubation of the crystalline material in TMB solution under aerobic condition showed a deep coloration within minutes for CeFtn$^{(pos)}$/CeFtn$^{(neg)}$ crystals, indicating catalytic activity within the protein matrix (Figure 5.55A). To prove the formation of the reaction product, UV-Vis absorption spectrum of a thin crystal layer was measured (Figure 5.55C). The same spectrum as in solution was obtained with an absorption maximum at 650 nm, assigned to the oxidized TMB. Even a reaction was observed for crystals composed of one protein container loaded with CeO_2 (Appendix, Figure 8.20). Interestingly, the turnover was observed more quickly in the nanoparticle superlattice compared to the solution. Due to the confined space of the protein matrix, CeO_2 nanoparticles are concentrated in the crystal resulting in a visible concentration of the blue colored product. Peroxidase-like activity was also observed for all three nanoparticle superlattices (Appendix, Figure 8.21). Moreover, an assembly of CeFtn$^{(pos)}$/CeFtn$^{(neg)}$ crystals was incubated for 2 days with TMB and the product concentration in the supernatant was measured. The crystals were stable and enabled several cycles of catalytic turnover with high activity remaining in each cycle (Appendix, Figure 8.22). The nanoparticle superlattice was not inactivated by the substrates. As a step towards multifunctional materials, the activity of FeFtn$^{(neg)}$ for the oxidation of TMB was also studied. Only in presence of hydrogen peroxide the reaction occurred, indicating a peroxidase-like activity but no oxidase-like activity (Appendix, Figure 8.23). A similar reaction rate as for CeO_2 loaded protein composites was obtained (Appendix, Figure 8.24 and Table 8.3). Thus, the reaction rate is mainly limited by the protein container pores. As expected, experiments with the iron oxide nanoparticle superlattice showed the same results like in solution (Appendix, Figure 8.25). With hydrogen peroxide, a deep blue coloration was observed. Without hydrogen peroxide, no reaction takes place and the crystal remained unchanged.

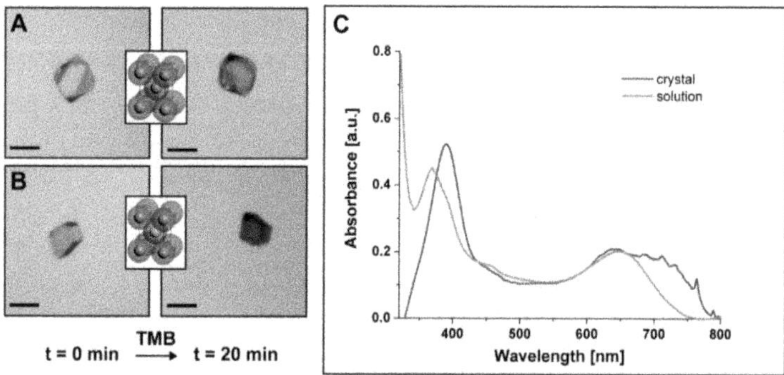

Figure 5.55: Oxidase-like activity of CeO_2 loaded protein crystals. (A) CeFtn$^{(pos)}$/CeFtn$^{(neg)}$ crystals before and after incubation with TMB for 20 minutes. (B) CeFtn$^{(pos)}$/FeFtn$^{(neg)}$ crystals before and after incubation with TMB for 20 minutes. (C) Absorption spectra of a thin layer of CeFtn$^{(pos)}$/CeFtn$^{(neg)}$ crystal after the reaction with TMB and of CeFtn$^{(pos)}$ in solution.

The oxidase-like activity of a binary CeFtn$^{(pos)}$/FeFtn$^{(neg)}$ crystal was studied (Figure 5.55B). Again, a blue coloration was observed so that iron oxide nanoparticles did not hamper the reactivity of the cerium oxide nanoparticles. Consequently, the combination of two different nanoparticles in the protein scaffold does not influence the activity of a single nanoparticle. Ideally, the protein crystal can be used as a filter for different substrates because of the high porosity. Hence, the oxidase-like activity of CeFtn$^{(pos)}$/CeFtn$^{(neg)}$ crystals was studied for several dyes with different sizes (Appendix, Figure 8.26-Figure 8.30). For all dyes, a coloration of the crystals was observed indicating an oxidation of the dye. Interestingly, most of the substrates are slightly larger in thickness than the container pore size (4 Å vs. TMB 5 Å). However the incorporation of large molecules such as organometallic complexes into the ferritin cavity was already observed due to the flexibility of the protein container pores.[153,292] Nevertheless, a filter effect can occur by choosing larger substrates or further engineering of the crystal lattice channels or container pores.

Since iron oxide nanoparticles possess similar catalytic properties as cerium oxide nanoparticles, a cascade reaction where individual reactions are carried out by each nanoparticle was difficult to find. However, originally, iron oxide nanoparticles were synthesized to obtain magnetic properties and not because of their interesting catalytic activity. Therefore, the magnetic properties of FeFtn$^{(neg)}$ were investigated in solution. Figure 5.56A shows the magnetization curve with a Langevin fit. The iron oxide nanoparticles possessed superparamagnetic behavior. From the fit, a saturation magnetization M_s of 30.125 ± 0.44 emu g^{-1} was determined. Compared to wildtype horse-spleen ferritin loaded with cobalt-doped iron oxide nanoparticles (M_s = 48.6 emu g^{-1}), a lower magnetic saturation was observed.[293] Moreover, iron oxide nanoparticle-loaded protein crystals were able to be manipulated inside a water drop by using a strong neodymium magnet, indicating magnetic properties of the nanoparticle superlattice (Figure 5.56B and C). Finally, CeFtn$^{(pos)}$/FeFtn$^{(neg)}$ crystals were used to perform the oxidation of TMB and subsequently moved in the drop by the magnet (Figure 5.56D and E). Thus, a combination of two nanoparticles is possible without loss of the properties of the single nanoparticle.

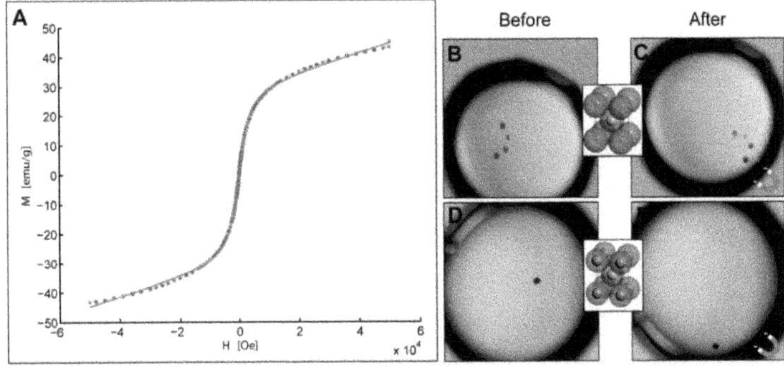

Figure 5.56: Magnetic properties of FeFtn$^{(neg)}$. (A) Magnetization with Langevin function fitted curve. Optical microscope image of eFtn$^{(pos)}$/FeFtn$^{(neg)}$ crystals (B) before and (C) after treatment with a magnet. Optical microscope image of CeFtn$^{(pos)}$/FeFtn$^{(neg)}$ crystal (D) before and (E) after treatment with a magnet.

In addition to the metal oxide nanoparticles, the catalytic activity of AuNPs was also studied. As a model system the reduction of 4-nitrophenol (4-NP) with $NaBH_4$ as hydride source was chosen. Here, the progress of the reaction can be monitored by UV-Vis spectroscopy: The catalytic turnover can be detected by the decrease of the absorption maximum of 4-NP at 400 nm. Figure 5.57A shows the reduction of 4-NP with free AuNP for a time interval of 30 minutes.

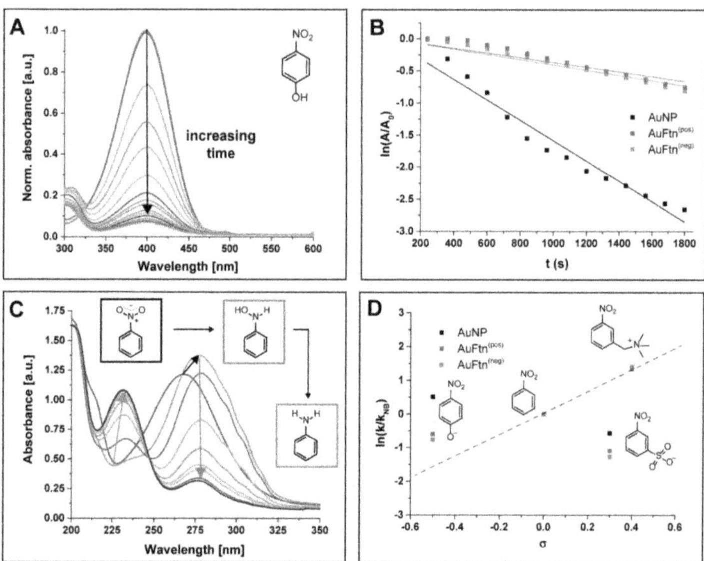

Figure 5.57: Catalytic reduction of nitroarenes. (A) Absorption spectra of 4-NP reduction with free AuNP for 30 minutes in 2 min intervals. (B) Determination of reaction rate for 4-NP reduction for free AuNP (black), AuFtn$^{(pos)}$ (blue) and AuFtn$^{(neg)}$ (red). (C) Absorption spectra of NB with free AuNP for 45 minutes in 3 min intervals. (D) Hammett plot for the reduction of substituted benzene derivates with AuNP (black), AuFtn$^{(pos)}$ (blue) and AuFtn$^{(neg)}$ (red).

The absorption maximum decreased over time, caused by the conversion of 4-NP to 4-aminophenol (4-AP). Additionally, a new absorption maximum of 4-AP at 300 nm emerged. The reduction of 4-NP is a reaction of first order, where the reaction rate is directly proportional to the concentration of 4-NP. Moreover, the reaction rate is independent on the starting concentration and only depends on the concentration of the catalyst. The reaction rate was determined for AuNP, AuFtn$^{(pos)}$ and AuFtn$^{(neg)}$. Before the reaction was examined, the AuNP samples were adjusted to a similar concentration (Appendix, Figure 8.31). To determine the reaction rate, ln(A/A$_0$) was plotted against reaction time (Figure 5.57B). As expected, the reaction rate was higher for the reaction with the free AuNP (0.00159 s^{-1}) compared to AuFtn$^{(pos)}$ (0.000367 s^{-1}) and AuFtn$^{(neg)}$ (0.000406 s^{-1}). Again 4-NP must pass through the protein container pores to access the AuNP for the catalytic reaction resulting in a lower reaction rate. To examine the influence of the substrate for the reaction, three additional substrates were studied with different functional groups: nitrobenzene (NB), 3-nitrobenzesulfonate (NBS) and (3-nitrobenzyl)trimethylammonium (NTA). Figure 5.57C shows the reduction of NB, which is representative for the three additional substrates. In contrast to

4-NP, the reaction progress cannot be monitored directly by the decrease of the absorption maximum of NB. In literature, the formation of hydroxylamine as an intermediate is described.[294] However, the formation of the intermediate is fast compared to the subsequent reduction of the intermediate to the final aminoarene. Therefore, the decrease of intermediate absorption maximum can be used to determine reactions rate. In detail, at the beginning of the reaction, the absorption maximum of NB at 269 nm was visible. Immediately, the intermediate (Figure 5.57C, red box) was formed with an absorption maximum at 280 nm. Full conversion of NB was directly observed because the peak at 269 nm disappeared (Figure 5.57C, black arrow). During the reaction, a decrease of the signal at 280 nm occurred while a new peak at 230 nm emerged, assigned to the final aminoarene (Figure 5.57C, green box). The same observation was made for NBS and NTA (Appendix, Figure 8.33 and Figure 8.34). Reaction rates were examined similar as for 4-NP and are summarized in Table 5.13.

Table 5.13: Reaction rates of the reduction of substituted benzene derivates with different AuNP samples.

Substrate	AuNP	AuFtn$^{(pos)}$	AuFtn$^{(neg)}$
4-NP	1.59×10^{-3} s^{-1}	3.67×10^{-4} s^{-1}	4.06×10^{-4} s^{-1}
NB	9.56×10^{-4} s^{-1}	6.63×10^{-4} s^{-1}	8.61×10^{-4} s^{-1}
NTA	3.83×10^{-3} s^{-1}	2.55×10^{-3} s^{-1}	3.51×10^{-3} s^{-1}
NBS	5.45×10^{-4} s^{-1}	2.20×10^{-4} s^{-1}	2.40×10^{-4} s^{-1}

To compare the reactions rate of the different AuNP samples, a Hammett plot was performed. The idea of the Hammett plot is that the change in free energy of activation is proportional to the change of Gibbs free energy for reactions with aromatic reactants only differing in the type of substitution of the arene ring.[295] Thus, a substituent constant σ for each substituent is defined. The relation between substituent constant and reaction rate is described by equation 4.1:

$$ln\frac{k}{k_0} = \sigma p \quad (4.1)$$

where p is the reaction constant, k_0 the reaction rate of the unsubstituted benzene derivate and k the reaction rate of the substituted benzene derivate. Ideally, a linear curve with p as slope would be obtained. However, no linear curve was obtained (Figure 5.57D). Interestingly, for negatively charged AuNPs in CCMV a similar behavior was observed.[296] There, the difference from the Hammett plot was explained due to the negatively charged AuNPs, which influence the reactivity of the substrates. It seemed that the positively charged AuNP in our system had the same effect on the reactivity. Moreover, a comparison of the different AuNP samples revealed further differences. The $\ln(k/k_0)$ values for both AuNP-loaded protein containers are the same, so that the outer surface did not influence the reactivity. However, compared with the free AuNPs, the same value was obtained for NTA but lower values for 4-NP and NBS. NTA is positively charged but 4-NP and NBS are negatively charged. Due to the negatively charged pores in the ferritin protein container,[163] the diffusion of the negatively charged substrates was possibly hampered, resulting in an inferior reaction rate.

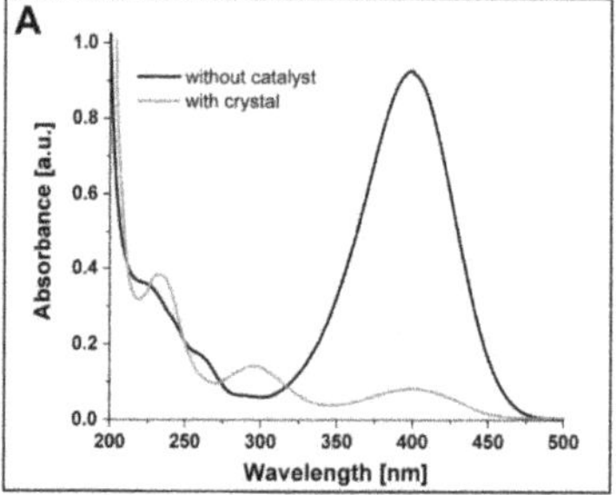

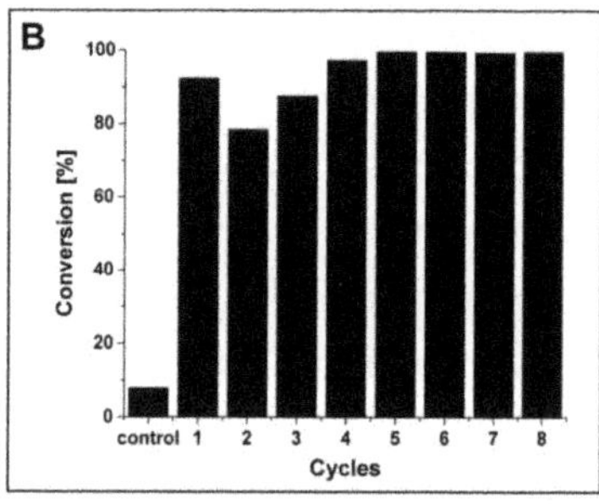

Figure 5.58: Catalytic reduction of 4-NP in the protein crystal. (A) Absorbance spectra of 4-NP without catalyst (black) and with AuFtn(pos)/eFtn(neg) crystals as catalyst (red). (B) Several catalytic reactions with the same crystals.

Finally, the catalytic reactivity of AuFtn(pos)/eFtn(neg) crystals was studied by the reduction of 4-NP. The supernatant was measured using UV-Vis spectroscopy and a decrease of the absorption maximum was examined, indicating the successful reduction of 4-NP (Figure 5.58A). Similar to the catalysis in solution, the reaction rate was determined (Appendix, Figure 8.36). A reaction rate of $1.76 \times 10^{-3}\,s^{-1}$ was determined, which was approximately half of the reaction rate of the AuNP-loaded protein container in solution. Nevertheless, a comparison is difficult because of the unknown amount of catalyst in the crystals. Moreover, the same crystals were used several times without observing a loss of the catalytic activity (Figure 5.58B).

In summary, nanoparticle-loaded protein crystals showed that the assembly of building blocks in the protein crystal led to an interaction between different cargos in the material, resulting in new properties. Moreover, the nanomaterial can be used as heterogonous catalyst for oxidation as well as reduction reactions. The combination of two different nanoparticles did not hamper the catalytic activity of the single nanoparticle. Finally, the combination of a catalytic active nanoparticle and a magnetic nanoparticle was possible to obtain a binary nanomaterial containing the properties of both nanoparticles.

6 Summary and perspective

In conclusion, nanoparticle-loaded protein containers were utilized as building blocks for the construction of biohybrid materials with unique properties. In addition to the direct synthesis of metal oxide nanoparticles inside the protein containers, a novel encapsulation method was presented by encapsulation of pre-synthesized AuNP via dis- and reassembly of the protein container.

For this purpose, AuNP and AgNP with a size between 3 and 4 nm were synthesized. Ligand exchange reactions were performed to obtain positively charged water-soluble nanoparticles suitable for the encapsulation inside the protein containers. Towards this goal, a new ligand shell system was introduced based on a mixture of tertiary and quaternary ammonium ligands. This mixed ligand shell was further characterized by SERS, NMR and TGA to determine the exact ligand shell composition. Moreover, the stability of AuNP and AgNP with different ligand shells was investigated and a higher colloidal stability of the AuNP compared to the AgNP was observed.

The dis- and reassembly behavior of both protein variants ($Ftn^{(pos)}$ and $Ftn^{(neg)}$) in an acidic and chaotropic condition was studied. On the one hand, the reassembly of $Ftn^{(pos)}$ was only successful in the chaotropic condition. On the other hand, $Ftn^{(neg)}$ reassembled to form the whole protein container in both conditions. Importantly, the crystallization to construct the protein scaffold was feasible for all samples. However, the reassembled $Ftn^{(neg)}$ sample in the acidic condition required an additional purification step.

Encapsulation of AuNPs was performed for both protein variants. By using AuNPs with different ligand shell thickness, no alteration of the protein container size was observed. Instead, smaller AuNPs were encapsulated with increased ligand shell thickness indicating a size exclusion effect during the encapsulation process. Interestingly, the total size (core + ligand) of the encapsulated AuNPs equals the size of the synthesized metal oxide nanoparticles inside the protein container cavity. Furthermore, a salt screening was performed to find the optimal ionic strength for the highest encapsulation efficiency. Importantly, the same ionic strength led to the best results for both protein container variants with the chaotropic encapsulation condition. Thus, the encapsulation is mostly driven by electrostatic interaction of the inner cavity with the AuNP. For the acidic condition, a lower ionic strength is required because of the different charge of the AuNP during the encapsulation. Separation of free AuNPs and nanoparticle filled protein containers was performed with IEC. Here, the amount of nanoparticle filled protein containers was increased up to 98% for $Ftn^{(pos)}$ and 92% for $Ftn^{(neg)}$ by increasing the amount of AuNPs to two-fold and 4.5-fold excess, respectively. To improve the purity of the sample, a further purification by SEC was feasible. Encapsulation of AgNPs was not possible due to the lower colloidal stability of AgNPs. Further optimization of the AgNPs in terms of colloidal stability is required to ensure the synthesis of these nanoparticle-loaded building block. To prove the universal usage of the dis- and reassembly approach, encapsulation of other types of nanoparticles should be performed.

The assembly of the AuNP-loaded protein containers by protein crystallization into highly ordered structures was realized. The crystalline structures were investigated by optical microscopy, SEM and SAXS. Importantly, the structure of the nanoparticle superlattice is determined by the protein scaffold. As an alternative to AgNP-loaded protein containers, dye-loaded protein containers were synthesized. AuNP-loaded protein containers were combined with dye-loaded protein containers as well as metal oxide loaded protein containers to obtain several binary nanomaterials. By alteration of the cargo inside the protein-based building blocks, the properties of the nanomaterial are tunable for the application as catalyst or as an optically active material. For the investigation of material properties usually a large amount of material is required. However, the crystallization with the hanging drop vapor diffusion method yields only small amounts of crystals. Therefore, batch crystallization was optimized successfully to obtain larger amounts of the protein crystals. Upscaling up to 50 µL was performed and can possibly be increased even further.

Finally, the optical and catalytic properties of the nanoparticle-loaded protein crystals were investigated. The optical properties of different AuNP and dye-loaded crystals were studied with a confocal microscope. On the one hand, a $eFtn^{(pos)}/RhFtn^{(neg)}$ crystal showed strong fluorescence. On the other hand, a $AuFtn^{(pos)}/RhFtn^{(neg)}$ crystal showed no fluorescence because of interactions between the dye and the AuNP, resulting in a quenching of the fluorescence. Importantly, the distances in the crystal are appropriate to observe interparticle interactions between the cargos of the building blocks. Further experiments by altering the emission maximum of the incorporated dye molecule could be interesting to tune and investigate the interactions between the building blocks. For the application as SERS substrate, the AuNP-loaded protein crystals were not suitable. Nevertheless, the weak enhancement of the dried AuNP drop showed the potential for the enhancement of Raman signals. Until now the focus was laid to three-dimensional nanoparticle superlattices. The construction of thin layers of the nanoparticle-loaded protein containers could also lead to interesting properties.

In addition to the optical properties, the catalytic properties were studied. The oxidase-like and peroxidase-like activity of the cerium oxide nanoparticles were investigated by the oxidation of TMB in solution and *in crystallo.* Moreover, the combination of cerium oxide and iron oxide nanoparticles in the protein crystal did not hamper the catalytic activity of the cerium oxide nanoparticle. Importantly, $CeFtn^{(pos)}/FeFtn^{(neg)}$ showed catalytic and magnetic properties, resulting in a multifunctional material. Besides AuNP-loaded protein crystals were utilized for the reduction of different nitroarenes. Interestingly, a combination of optically active gold and magnetic iron oxide nanoparticles could lead to novel magneto-optic nanomaterials, which can be utilized in magneto-electronics.

Although different catalytic reactions (oxidation, reduction) are feasible, it was not possible to find a cascade reaction, where two consecutive reactions will be performed to yield a product. The executed reactions were carried out in aqueous solutions. First stability experiments in organic solvents and with higher temperature showed an increased stability of the cross-linked crystals compared to untreated crystals. The stability of the crystals was confirmed macroscopically under an optical microscope. Further structural studies with SAXS of crystal ensemble or single crystal would be required to investigate the microscopic influence of organic solvents and temperature on the structure. Importantly, an increased stability in harsher conditions would enhance the field of applications for catalytic reactions and so the probability to find a cascade reaction for the novel nanomaterial.

The focus in this work was put on at binary protein crystals composed of two oppositely charged ferritin protein container. Importantly, the crystal composition and lattice parameters are easily tunable by alteration of the crystallization. Although several mutations were introduced to supercharge the protein containers outer surface and to ensure binary crystallization, several mutations are not involved in the crystal contact formation. A detail study of the crystallization process would be interesting to understand and to tune the formation of the crystal lattice in a more reasonable way.

Therefore, several experiments can be performed such as the deletion of the mutations currently not participating in crystal contacts, or the alteration of the existing crystal contacts to observe the consequence for crystal formation.

In addition to the ferritin protein container, the encapsulin protein container is recently used in our research group. Encapsulin has an outer diameter of 24 nm and an inner diameter of 20 nm. As an extension, the crystallization of different sized protein containers would enhance the library of different structures for the creation of nanomaterials. This modularity can enable the formation of different crystal structures in AB, AB_2 or other compositions. Currently, a crystallization condition was found for the crystallization of $Ftn^{(pos)}$ and wild-type encapsulin. Control experiments with both proteins alone resulted in no crystal formation. Unfortunately, no diffraction was measured from the crystals at the synchrotron. As an alternative, TEM measurement of the crystals can be performed to determine the structure. However, the only difficulty is to obtain thin layers of the crystals so that the electron beam can pass through the sample. The larger encapsulin protein container will enable the encapsulation of larger plasmonic nanoparticles with increased plasmonic coupling. Combination with dye-loaded or magnetic iron oxide-loaded ferritin protein containers can lead to novel optical materials with unique properties due to short interparticle coupling between the building blocks.

7 Experimental Part

7.1 General

All syntheses were carried out in ultrapure water from the *Purelab Plus* system manufactured by *ELGA LabWater* (*Celle*). The electrical conductivity was 0.055 µS/mL and the water was filtered through 0.22 µm membrane filters (*Merck*). All glassware and magnetic stir bars were cleaned with aqua regia and rinsed several times with ultrapure water to ensure removal of adsorbents. Particle solutions were centrifuged with 15 mL or 50 mL polystyrene tubes (*Carl Roth*) with an *Eppendorf 5810R* centrifuge or small volumes in 1.5 mL *Eppendorf* tubes with a *Heraeus Fresco 21* microcentrifuge from *Thermo Scientific*. Competent cells and cryo cultures were stored at -80 °C. All cell pellets and nucleic acids were stored in the freezer at -20°C. *Amicon Ultra-15*, *Amicon Ultra-4* or *Amicon Ultra-0.5* centrifugal filter units from *Merck* were used for the dialysis (including buffer exchange, washing and concentration) of particle solutions and protein containers. A molecular cut-off (MWCO) of 30 kDa was used. Purification of protein containers by ion exchange or size exclusion chromatography was performed with an *Äkta pure 25* system from *GE Healthcare Life Science*. Concentrations of the protein solutions were determined with a *NanoDrop 2000* spectrophotometer from *Thermo Scientific*. If not stated otherwise, molar concentrations of the proteins always referred to assembled containers. Particle solutions and purified protein solutions were stored in the dark at 4°C.

7.2 Chemicals

All chemicals were purchased from commercial sources (see Table of chemicals on page 175).

7.3 E-coli strains

Table 7.1: List of used strains.

Strain	Genotype	Supplier
BL21(DE3)	*E. coli* B F- *dcm ompT hsdS*(rB- mB-) *gal* λ(DE3)	Agilent
DH5α	F^- Φ80*lac*ZΔM15 Δ(*lac*ZYA-*arg*F) U169 *rec*A1 *end*A1 *hsd*R17(rk^-, mk^+) *pho*A *sup*E44 *thi*-1 *gyr*A96 *rel*A1 λ	Invitrogen

The cDNA of the ferritin variants were cloned into a pET-22b(+) expression plasmid (*Novagen*).

7.4 Used buffers

All buffers were produced in ultrapure water and subsequently filtered through 0.22 µm membrane filters (*Merck*).

Table 7.2: Buffers used for the purification of the protein variants, the synthesis of nanoparticles and dis- and reassembly of the protein containers.

Buffer	Composition	pH value
A	50 mM TRIS + 0.15 M NaCl	7.5
B	50 mM TRIS + 0.30 M NaCl	7.5
C	50 mM TRIS + 1.00 M NaCl	7.5
D	50 mM TRIS + 1.50 M NaCl	7.5
E	50 mM MES + 0.50 M NaCl	6.0
F	200 mM HEPES + 0.2 M NaCl	8.6
G	10 mM phosphate + 0.05 M NaCl	2.0
H	20 mM phosphate + 0.05 M NaCl	7.4
I	20 mM TRIS + 0.15 M NaCl	7.5
J	20 mM acetate + 0.05 M NaCl	5.0

7.5 Analytics

7.5.1 Transmission electron microscopy

Transmission electron microscopy (TEM) measurements were performed on a *Zeiss Libra 200* FE transmission electron microscope with energy filters and operated at 200 kV. Samples with nanoparticles were prepared by drying 2 µL of the sample on Formvar carbon-coated copper grids, 400 mesh (TED Pella, 01814-F). Samples with protein containers were stained with uranyl acetate (2% stock solution) to make the protein containers visible via negative stain. To this end, the grid was incubated face down on a 10 µL droplet of sample for 60 s. Protein concentrations of 0.1 to 0.2 mg/mL for $Ftn^{(pos)}$ and 0.2 to 0.3 mg/mL for $Ftn^{(neg)}$ were used. The grid was washed three times in water, followed by one wash and 45 s incubation on uranyl acetate droplets. Finally, the grid was blotted and dried under air. Images were analyzed with the ImageJ software.[297] The size of the protein containers was manually determined by analyzing 200 particles per sample. In the program, a circle was used with a size that fitted the whole protein container. The area of that circle was measured to determine the size. AuNP size was determined by converting the images into binary images with the threshold function, followed by automatic counting and area determination.

7.5.2 Scanning electron microscopy

Scanning electron microscopy (SEM) measurements were performed by *Robert Schön* (University of Hamburg, Germany) on a *Zeiss Leo1550* scanning electron microscope and operated at 2.00 kV. Energy-dispersive electron X-ray spectroscopy (EDX) was carried out with an EDX detector (SDD 100 mm^2). EDX maps were prepared with the *Oxford Instruments* software. For the sample preparation, protein crystals were washed with water, transferred into a water drop on a silicon waiver and dried under air.

7.5.3 Dynamic light scattering

Dynamic light scattering (DLS) measurements were carried out on a *Zetasizer Nano S* from *Malvern Instrument* equipped with HeNe laser (wavelength 633 nm) at a backscattering angle of 173°. All samples were measured at 20°C in the storage buffer for protein container samples and in H_2O MilliQ or DCM for AuNP samples. Protein concentration was adjusted to 4 mg/mL. AuNP samples were diluted until the plasmonic peak reached an intensity below 0.5 AU. Aqueous solutions were usually measured in disposables cuvettes (PMMA, semi-micro) and organic solutions in quartz cuvettes. Prior to the measurement, the samples were equilibrated for 120 s. Data were analyzed with *Malvern DTS v. 7.02* software by cumulant analysis to determine the hydrodynamic radius. At least three runs were averaged.

7.5.4 Zeta potential

Zeta potential (ζ-pot) measurements were conducted with a *Zetasizer Nano S* from *Malvern Instrument* equipped with HeNe laser (wavelength 633 nm) at 20°C in micro zeta disposable capillary cells (PMMA, DTS1070, *Malvern Instrument*). Prior to the measurement, the samples were equilibrated for 120 s. Protein concentration was 1 mg/mL. AuNP concentration was the same as for DLS. Data analysis was performed with the Smoluchowski approximation to convert electrophoretic mobility to the ζ-pot. At least three runs were averaged.

7.5.5 Surface enhanced Raman spectroscopy

Surface enhanced Raman spectroscopy (SERS) spectra were measured by *Sophia Peter* (Institute of Inorganic Chemistry, RWTH Aachen University, Germany) with a *Horiba Jobin Yvon* Raman spectrometer at a laser wavelength of 633 nm and 180° backscattering arrangement with *Olympus* objectives at 50-fold magnification. The spectral resolution was 7 cm^{-1} and the power on the sample was 12 mW. For sample preparation, 2 µL of sample were dried on a gold (thickness 100 nm) coated silicon wafer.

7.5.6 Atomic absorption spectroscopy

Atomic absorption spectroscopy (AAS) measurements were performed by *Björn Faßbender* (Institute of Inorganic Chemistry, RWTH Aachen University, Germany) on an *AA-6200* atomic absorption spectrophotometer from *Shimadzu* at a detection wavelength of 242.8 nm. The samples were treated with aqua regia prior to measurement.

7.5.7 Nuclear magnetic resonance spectroscopy

^{1}H-nuclear magnetic resonance (^{1}H-NMR) spectra were recorded by *Dr. Gerhard Fink* (Institute of Inorganic Chemistry, RWTH Aachen University, Germany) on an *Avant II spectrometer* (Bruker) at 400 MHz with D_2O as internal standard. In addition, the WET-technique (Water Suppression Enhanced through T_1-Effects) was performed to suppress the solvent signal and amplify the signals of the ligands.[243,298]

AuNP were precipitated under vacuum overnight, which resulted in a dry pellet to get a water-free sample. The precipitated AuNPs were dissolved with 485 µL of D_2O and 15 µL (50 mg/mL) DTT was added. Ligand exchange with DTT occurred and the AuNPs precipitated. The sample was centrifugated and NMR spectra of the supernatant composed of the mixed ligand shell and DTT were measured.

7.5.8 UV-Vis spectroscopy

Transmission spectra were recorded on a *JASCO V-630* spectrophotometer. Aqueous solutions were usually measured in disposables cuvettes (PMMA, semi-micro) and organic solutions in quartz cuvettes.

Emission spectra of the crystals were measured by *Roman Kusterer* (Institute of Physical Chemistry, University of Hamburg, Germany) on a home-built confocal microscope.

7.5.9 Optical microscopy

Imaging of the crystal samples was performed with two different setups. Both setups had in common that the crystals were directly imaged in the manual plates. For imaging a *Zeiss Stemi SV 11* stereomicroscope equipped with a *Canon EON Rebel T1i* camera and later a *CrysCam™ Digital Microscope* (Dunn Labortechnik GmbH) was used. The scale was determined with a ruler.

7.5.10 X-ray diffraction

Powder X-ray diffraction (XRD) measurements were conducted at room temperature using a *Stoe STADI MP* diffractometer operating in transmission mode with Mo K_α radiation (λ=0.71 Å). The acquired powder XRD data were processed with the software *WinXPOW*.

Liquid samples were precipitated and dried under vacuum. For protein containers, 1% SDS solution and for commercially available nanoparticles buffers with high sodium chloride concentration were used. A foil was prepared with grease, the sample was placed in the middle and a second foil was placed on the top to trap the sample.

7.5.11 Small angle X-ray scattering

SAXS data were obtained from cross-linked crystals in mother liquor and transferred to a polyimide capillary. The capillary was mounted into a capillary holder and rotated to ensure data collection of the crystal ensemble and not of only few crystals. SAXS data were collected at an *Incoatec* X-ray source IµS with *Quazar Montel* optics in a q range from 0.00007 Å^{-1} - 0.25 Å^{-1} by *Dr. Andreas Meyer* (Institute of Physical Chemistry, Universität Hamburg, Germany). The wavelength was 1.546 Å. The scattering patterns were collected with a *Rayonix SX165* CCD-detector. Sample-to-detector distance was 1.6 m. The magnitude of the scattering vector q is given by $q = 4\pi \sin\theta/\lambda$, where 2θ is the scattering angle.

Unit cell parameters for each data set were determined using the scattering vector q for reflections (100), (001) and (200). The relation between the scattering angle θ and the scattering vector q ($q = 4\pi \sin\theta/\lambda$) were combined with the quadratic Bragg equation for tetragonal crystal systems to obtain equation 7-1:

$$q_{\mathrm{hkl}} = \frac{2\pi}{a}\sqrt{h^2 + k^2 + l^2\left(\frac{a}{c}\right)^2} \quad (7\text{-}1)$$

Cell parameter *a* was determined by using reflections (100) and (200):

$$a = \frac{2\pi}{q_{100}} \quad (7\text{-}2)$$

$$a = \frac{4\pi}{q_{200}} \quad (7\text{-}3)$$

Reflection (001) was used to determine cell parameter *c*:

$$c = \frac{2\pi}{q_{001}} \quad (7\text{-}4)$$

The *q* values for reflections (101) and (110) were calculated using the derived cell parameters:

$$q_{101} = \frac{2\pi}{a}\sqrt{1 + \left(\frac{a}{c}\right)^2} \quad (7\text{-}5)$$

$$q_{110} = \frac{2\pi\sqrt{2}}{a} \quad (7\text{-}6)$$

to be compared with the experimental values. Subsequently, the determined unit cell parameters from the SAXS experiment were used to simulate diffraction patterns with *PowderCell* (Federal Institute for Material Research and Testing, Germany). Crystals composed of an empty protein container and an AuNP filled protein container form a primitive tetragonal lattice (space group P4/mmm, No. 123). For the simulation, a tenth of the lattice parameters were used and the peak position was corrected during the conversion from scattering angle to scattering vector. Simulated diffraction pattern were plotted together with experimental data in *Origin* (Origin Lab Corporation).

7.5.12 Thermogravimetric analysis

Thermogravimetric analysis (TGA) measurements and data evaluation were performed by *Brigitte Jansen* (Institute of Inorganic Chemistry, RWTH Aachen University, Germany) using a *Mettler STAR^e SW 9.20* instrument (*Mettler Toledo*). AuNP solution was dried in standard alumina pans (70 uL) until an amount of 3.0 mg was reached. The samples were measured in the temperature range from 25.0°C to 400°C at 2°C/min under a nitrogen flow of 60 mL/min.

The ligand footprint (FP) can be calculated from the mass loss during the TGA measurements.[238,246,299] Due to different starting values for the TGA measurements, an effective mass loss according to equation 7-7 was calculated:

$$\Delta m_{\mathrm{eff}} = \frac{\Delta m_{\mathrm{AuNP}}}{\Delta m_{\mathrm{Ligand}}} \quad (7\text{-}7)$$

Δm_{eff}	effective mass loss
Δm_{AuNP}	mass loss from the AuNP
Δm_{Ligand}	mass loss from the ligands

From the effective mass loss, the number of ligands n_{C2mix} and the amount of AuNP n_{AuNP} can be determined according to equations 7-8 and 7-9:

$$n_{\mathrm{C2mix}} = \frac{\Delta m_{\mathrm{eff}}}{M_{\mathrm{C2mix}}} \quad (7\text{-}8)$$

n_{C2mix}	amount of C2mix
M_{C2mix}	averaged molecular weight of both ligands (148.66 g mol^{-1})

$$n_{\mathrm{AuNP}} = \frac{n_{\mathrm{Au}}}{N_{\mathrm{Atom/NP}}} = \frac{\frac{(1-\Delta m_{\mathrm{eff}})}{M_{\mathrm{Au}}}}{N_{\mathrm{Atom/NP}}} \quad (7\text{-}9)$$

n_{AuNP}	amount of AuNP
n_{Au}	amount of gold after the weight loss in the TGA measurement
$N_{Atom/NP}$	number of gold atoms per AuNP (see chapter 7.8.3)
M_{Au}	molecular weight of gold

From these values, the number of ligands per AuNP $N_{Ligand/AuNP}$ can be calculated according to equation 7-10:

$$N_{\mathrm{Ligand/AuNP}} = \frac{n_{\mathrm{C2mix}}}{n_{\mathrm{AuNP}}} \quad (7\text{-}10)$$

Finally, the FP is obtained by equation 7-11:

$$FP = \frac{A_{\mathrm{AuNP}}}{N_{\mathrm{Ligand/AuNP}}} \quad (7\text{-}11)$$

FP	footprint in nm^2
A_{AuNP}	surface area of a spherical AuNP in nm^2

7.5.13 Circular dichroism spectroscopy

Circular dichroism (CD) spectra were recorded on a *JASCO CD-Photometer J-1100* in quartz cuvettes with a path length of 0.5 mm and a total volume of 80 µL. Data were measured from 190 nm to 300 nm in 0.2 nm intervals with a scanning speed of 200 nm/min. Protein solution concentration was 0.5 mg/mL. For macromolecules such as proteins, the mean residue molar ellipticity $[\Theta]_{MRW}$ according to equation 7-12 was used:

$$[\Theta]_{\mathrm{MRW}} = \frac{[\Theta]}{10\, c_r\, l} \quad (7\text{-}12)$$

$[\Theta]$	ellipticity in mdeg
c_r	mean residue molar concentration in mol^{-1} mL^{-1}
l	cell path in cm

Using the mean residue molecular concentration according to equation 7-13:

$$c_{\mathrm{r}} = \frac{1000\, n\, c_{\mathrm{g}}}{M_{\mathrm{r}}} \quad (7\text{-}13)$$

n	number of peptide bonds
c_g	protein concentration in g mL^{-1}
M_r	molecular weight of the protein container

7.5.14 Density functional theory calculations

All calculations were performed based on the Kohn-Sham density functional theory (DFT) as implemented in Gaussian 09 program suite.[227] The start geometry was optimized using the basis theorem MP2 6-311G++. All geometries were characterized as stationary points without imaginary frequencies.

7.5.15 Magnetic characterization

For sample preparation, 100 µL of FeFtn$^{(neg)}$ was prepared with an iron concentration of 0.5 mg/mL (concentration was determined by AAS). The magnetic characterization of the sample was performed by *Dr. Ioana Slabu* (Institut für Angewandte Medizintechnik, RWTH Aachen University, Germany) using a SQUID magnetometer (MPMS-XL, *Quantum Design*). The measurement was performed at room temperature, varying the magnetic field strength between 0 kOe and 30 kOe. The *Chantrell fitting* was used to determine the saturation magnetization value.[300]

7.6 Production and purification of the ferritin variants

7.6.1 Positively charged protein container

For precultures, single colonies of transformed *E. coli* BL21-Gold(DE3) cells were incubated overnight in 5 mL sterile LB-Miller medium supplemented with 150 µg/mL sodium ampicillin at 37°C and 250 rpm. 400 mL sterile LB-Miller medium supplemented with 150 µg/mL sodium ampicillin were inoculated with 4 mL of the preculture and incubated at 37 °C and 250 rpm. After reaching an OD_{600} of 0.2, the protein overexpression was induced with isopropyl β-D-1-thiogalactopyranoside (final concentration: 0.25 mM). The cultures were incubated at 37 °C for 5 h. Cells were harvested by centrifugation at 4°C and 5000 g for 20 minutes with a *Sorvall RC 6 Plus* centrifuge (*Thermo Scientific*). Pellets were washed with buffer C and stored at -20°C.

For purification, a pellet was thawed and resuspended in 10 mL of buffer D. The resuspended pellet was sonicated 4 x 1 minute on ice (amplitude: 60 %) with a *Vibra-Cell VCX-130* ultrasonic processor (*Sonics*). The suspension was centrifuged at 4°C and 14 000 g for 15 minutes. The supernatant was incubated with 25 mg (2.5 mg/mL) RNase at 37°C for 3.5 h. Afterwards the solution was heated to 65°C in a water bath for 10 minutes and centrifuged at 4°C and 14 000 g for 15 minutes. Saturated ammonium sulfate solution was added to the supernatant until a concentration of 70% of its saturation (23.3 mL) was reached. The solution was centrifuged at 4°C and 14 000 g for 15 minutes. Subsequently, the pellet was redissolved in 10 mL of buffer D. The solution was again precipitated with saturated ammonium sulfate solution and centrifuged as before. The pellet was redissolved in 50 ml of buffer E and filtered with a 0.22 µm syringe filter (TPP).

Subsequently, ion exchange and size exclusion chromatography were performed. For ion exchange chromatography, the sample was purified on a 5 mL *Sepharose HiTrap SP HP* cation exchange column (*GE Healthcare*) with an elution gradient from 0.5 M to 1.5 M NaCl (using buffer E and D). The $Ftn^{(pos)}$ containing fractions were collected and the protein solution was concentrated to 2 mL final volume. For size exclusion chromatography, the sample was purified on a *Hiload 16/600 Superdex 200 PG* size exclusion column (*GE Healthcare*) in buffer C. For crystallization experiments, the size exclusion step was repeated to achieve higher purity. As a last step, the sample was rebuffered at least five times with a centrifugal filter unit to the respective buffer for further experiments.

7.6.2 Negatively charged protein container

The production of Ftn$^{(neg)}$ was similar to Ftn$^{(pos)}$. The preculture was produced as described above. 200 mL sterile TB-Miller medium supplemented with 150 µg/mL sodium ampicillin were inoculated with 4 mL of the preculture and incubated at 37°C and 250 rpm. Protein overexpression was induced with isopropyl β-D-1-thiogalactopyranoside (final concentration: 0.25 mM) after an OD_{600} of above 0.6 was reached. The cultures were incubated at 18°C for 48 h. The cells were harvested similar to Ftn$^{(pos)}$, washed with buffer B and stored at -20°C.

The purification protocol was similar to Ftn$^{(pos)}$ with minor changes. The pellet was resuspended in 20 mL of buffer A. The resuspended pellet was sonicated for 8 x 1 min on ice. In contrast to Ftn$^{(pos)}$, the supernatant was not incubated with RNase. Heat precipitation was performed at 65°C in a water bath for 10 minutes and precipitation by ammonium sulfate was done twice as for Ftn$^{(pos)}$. The pellet was resuspended in 50 mL of buffer A and filtered with a 0.22 µm syringe filter (TPP).

The ion exchange was performed with a linear gradient from 0.15 M to 1.0 M NaCl on 5 mL *Sepharose HiTrap SP Q* cation exchange column (*GE Healthcare*). Gel filtration and further purification were the same as for Ftn$^{(pos)}$ but in buffer B.

7.7 Nanoparticle synthesis in the protein container

7.7.1 Synthesis of CeO_2 nanoparticles

The synthesis of CeO_2 in the cavity of the protein containers was the same for Ftn$^{(pos)}$ and Ftn$^{(neg)}$ with a minor change: Buffer B was used for the synthesis in Ftn$^{(neg)}$ and buffer C for the synthesis in Ftn$^{(pos)}$. Buffer and H_2O MilliQ were degassed by bubbling N_2 through the solutions for at least 15 minutes. The buffer was preheated in a round-bottom flask using an oil bath at 65°C for 20 minutes under N_2 (to maintain oxygen free environment) with constant stirring. Subsequently, 8 mg of protein were added to a total volume of 20 mL. 1.14 mL of H_2O_2 solution (15 mM) and $CeCl_3$ (30 mM) dissolved in deaerated H_2O MilliQ were added simultaneously using *Perfusor* compact S syringe pumps (B. Braun) in 1 h. Up to 2225 Ce(II) ions were injected per ferritin cage at a flow rate of 37 ions per minute per ferritin cage. After the addition was completed, the solution was equilibrated at 65°C for 15 minutes. 1.20 mL of EDTA (0.5 M stock solution) was added to chelate residual cerium ions and the solution was stirred for another 15 minutes.

For purification, the solution was centrifuged at 4°C and 14 000 g for 15 minutes and washed with H_2O MilliQ. Before the sample was further purified by size exclusion, the sample was centrifuged at least five times at 4°C and 20 000 g for 10 minutes. After the size exclusion, a sucrose gradient centrifugation was performed to separate loaded ferritin cages from empty or sparsely loaded containers. 2 ml of the sample were layered at the top of a stepwise sucrose gradient (3 mL of 20%, 7 mL of 40%, 8 mL of 70%) in 20 mL ultracentrifugation tubes. The sample was centrifuged with an *Optima XPN-80* ultracentrifuge (*Beckman Coulter*) at 4°C and 193 000 g for 3 h. After centrifugation, beginning at the top, 1 mL fractions were separated using a pipette. UV absorbance was measured with a *NanoDrop 2000* spectrophotometer. By comparison of the absorbance at 280 nm and 322 nm only fractions with well-loaded ferritin cages were chosen. The fractions were washed at least five times with water to remove the sucrose. Subsequently, the sample was rebuffered and the concentration was measured with the Bradford method.[301] Nanoparticle encapsulation was confirmed by TEM.

7.7.2 Synthesis of FeO_x nanoparticles

The synthesis of FeO_x nanoparticles was carried out with some modification. For $Ftn^{(neg)}$ the buffer was changed from buffer B to F. In detail, 25 mg of protein was dissolved in 3.65 mL H_2O MilliQ and mixed with 3.65 mL of buffer F to a total volume of 7.30 mL. After 15min, H_2O_2 (25 mM) and $(NH_4)_2Fe(SO_4)_2$ (75 mM) dissolved in deaerated H_2O MilliQ were added in equal volumes using *Perfusor* compact S syringe pumps (B. Braun). Up to 3500 Fe(II) ions were injected per ferritin cage at a flow rate of 9 ions per minute per ferritin cage. Once the addition was complete, the solution was kept in the water bath for another 15 min, and finally 400 µL sodium citrate (300 mM stock solution) were added to chelate residual iron ions. After additional 15 min at 65°C, the sample was centrifuged several times before injected on a *HiLoad 16/600 Superdex 200 PG* size exclusion column. Subsequently, the sample was rebuffered and the concentration was measured with the Bradford method.[301] Nanoparticle encapsulation was confirmed by TEM.

7.8 Gold nanoparticle synthesis and functionalization

For the synthesis of AuNPs, two different approaches were performed and optimized to obtain the desired particles size.[76,223] Both syntheses have in common that the AuNPs were obtained by reduction of tetrachloroauric acid and subsequent stabilization with ligands. Furthermore, the nanoparticles are dissolved in organic solvent to simplify the ligand exchange. All samples were analyzed by TEM, DLS and UV-Vis.

7.8.1 Leff synthesis

20.0 mg (0.050 mmol) of tetrachloroauric acid trihydrate ($HAuCl_4 \cdot 3\ H_2O$) was dissolved in 5 mL H_2O MilliQ and mixed with the necessary amount of dodecylamine (DDA, see Table 7.3) in 5 mL toluene in a beaker. The nanoparticle size was optimized by variation of the gold to ligand ratio. The mixture was stirred rigorously and 37.8 mg (0.76 mmol) $NaBH_4$ dissolved in 5 mL H_2O MilliQ was added dropwise in one minute. The beaker was sealed and stirred for 5 hours. For purification, the reddish organic phase was collected, and 40 mL of methanol was added to precipitate the AuNPs. The black precipitate was isolated by centrifugation (6000 G, 8 min) and washed with methanol. The procedure was repeated, and the black precipitate was dissolved in 10 mL dichloromethane. The yield calculated from AAS was approximately 7%.

Table 7.3: Different gold to ligand ratio for the optimization of the AuNP size.

gold:ligand ratio	*n* (DDA)
1:8	0.50 mmol
1:9	0.55 mmol
1:10	0.60 mmol
1:11	0.65 mmol

7.8.2 Peng synthesis

For the synthesis a precursor solution containing 2 mL tetralin (1.94 g, 14.60 mmol), 2 mL oleylamine (1.63 g, 6.08 mmol) and 20 mg $HAuCl_4 \cdot 3\ H_2O$ (0.050 mmol) was prepared and stirred for 10 minutes under nitrogen flow yielding an orange solution. A second reducing solution containing 0.2 mL tetralin (1.46 mmol), 0.2 mL oleylamine (0.61 mmol) and 32.2 mg (0.10 mmol) tert-butylamine borane complex (TBAB) was mixed by sonication and injected into the precursor solution. Immediately, the color of the solution changed to dark red and the solution was stirred for 1 h under nitrogen atmosphere. The reaction was terminated by adding 40 mL acetone. The black precipitate was isolated by centrifugation (6000 G, 8 min) and washed with acetone. The procedure was repeated, and the black precipitate was dissolved in 10 mL dichloromethane. Nanoparticle size optimization was performed by variation of the temperature (25°C, 30°C and 35°C). The yield calculated from AAS was approximately 45%.

7.8.3 Calculation of gold nanoparticle concentration

The AuNP concentration is needed to calculate the number of ligands for the ligand exchange. The gold concentration was determined by AAS and the AuNP concentration can be calculated using the number of atoms per NP according to equation 7-14:

$$c(AuNP) = \frac{c(Au)}{N_{Atom/AuNP}} \quad (7\text{-}14)$$

c(AuNP)	AuNP concentration in mol/L
c(Au)	Au concentration in mol/L
$N_{Atom/AuNP}$	number of atoms per AuNP

The number of atoms per NP was calculated according to equation 7-15.

$$N_{Atom/NP} = \frac{\rho_{Au}\, V_{AuNP}\, N_A}{M_{Au}} \quad (7\text{-}15)$$

ρ_{Au}	density of gold (19.32 g cm^{-3})
V_{AuNP}	AuNP volume in cm^3
N_A	Avogadro constant ($6.022 \cdot 10^{23}$ mol^{-1})
M_{Au}	Au molar mass in g mol^{-1}

It was assumed that AuNPs have a spherical form. The diameter was determined by TEM.

7.8.4 Ligand exchange

For the ligand exchange, (2-mercaptoethyl)-N,N,N-trimethylammonium chloride ($C2^+$) and 2-(dimethylamino)-ethanethiol hydrochloride (C2) were dissolved in H_2O MilliQ in equal amounts to create two different ligand solutions. These two solutions were mixed depending which ligand shell composition was desired and added in equal volume to the nanoparticle solution in a falcon tube. Routinely, a 20-fold excess was used as referred to the maximum number of ligands on the nanoparticle surface. For the ligand exchange with only one ligand the same procedure was performed. The mixture was shaken rigorously and a color change from the organic to the aqueous phase was observed. The sample was incubated at room temperature overnight. The aqueous phase was collected and washed five times with H_2O MilliQ using a centrifugal filter unit to remove excess of the ligands. Finally, the sample was concentrated to 1 mL. The ligand shell was analyzed with NMR, SERS and TGA.

7.9 Silver nanoparticle synthesis

For the synthesis of AgNPs, two different approaches were performed and optimized to obtain the desired particles size.[252,255] Both syntheses have in common that AgNPs were obtained by thermal deposition of silver ions in the presence of a capping ligand. Furthermore, the nanoparticles are dissolved in organic solvent to simplify the ligand exchange. All samples were analyzed by TEM, DLS and UV-Vis.

7.9.1 Yamamoto synthesis

The synthesis contained two steps. Firstly, 2.29 g (10.0 mmol) myristic acid in 30 mL MeOH was mixed with 0.46 g (11.5 mmol) NaOH in 10 mL H_2O MilliQ. By centrifugation (6000 g, 10 min) white flaky sodium myristate was obtained. 1.72 g (10.0 mmol) silver nitrate was dissolved in 10 mL H_2O MilliQ and added to the sodium myristate. The solution was shaken and grey silver myristate was collected by centrifugation. In a second step, 0.67 g (2.0 mmol) silver myristate was placed in a 100 mL flask and 5.8 mL NEt_3 (40.0 mmol) was added by a syringe. The mixture was stirred at 80°C for 30 minutes to produce a homogenous brown solution. 20 mL acetone was added yielding a dark yellow precipitate. The precipitate was washed three times with acetone by centrifugation (6000 g, 10 min). Finally, the particles were collected in toluene.

7.9.2 Hyeon synthesis

1.70 g (10 mmol) silver nitrate was dissolved in a mixture of oleylamine (0.5 mL, 1.5 mmol) and oleic acid (4.5 mL, 14.2 mmol). After degassing for 90 minutes, the solution was heated up to 180°C with different heating rates (see main text). By increasing the temperature, the color changes from grey to yellow and up to brown. After reaching the final temperature, the solution was let to cool down, before toluene (10 mL) and MeOH (40 mL) were added. The obtained product was collected by centrifugation (6000 g) and washed three times with MeOH (20 mL). The final product was dissolved in DCM.

7.9.3 Ligand exchange

AgNP concentration for the ligand exchange was calculated as for the AuNP. Moreover, the procedure for the ligand exchange reaction was the same with a minor change. After adding the ligand solution, the sample was incubated at 4°C in the dark.

7.10 Nanoparticle stability tests

The stability tests of the obtained nanoparticles after the ligand exchange were performed at RT and analyzed by UV-Vis spectroscopy. The protocol was in general the same for all tests. Nanoparticle solution was diluted to 300 µL of the respective condition so that all samples provide the same absorbance.

For ligand binding strength investigations, the nanoparticles were diluted to 0.1 M and 1 M DTT concentration, mixed thoroughly and measured overnight in 30 minutes intervals.

The influence of salt concentration was tested using different NaCl concentrations and 7 M Gua concentration and measured for 60 minutes.

The influence of the pH value was measured at different pH values (2.0, 7.0 and 12.0) for 120 minutes by UV-Vis spectroscopy. Additionally, ζ-pot measurements at different pH values (2.0, 6.0, 7.0, 10.0, 12.0) were performed.

7.11 Dis- and reassembly

The procedure was the same for both protein container variants. For the acidic approach 5 mg of the protein container was disassembled by adding 4 mL of buffer G (pH 2) and incubation for 4 h at room temperature. Reassembly was achieved by adding 20 mL of buffer H to adjust the pH back to 7.4 and incubation at 4°C overnight. For the chaotropic approach, the protein container was disassembled in 1 mL 7 M guanidinium hydrochloride solution. After incubation of 4 h at room temperature the sample was reassembled by dilution with H_2O MilliQ to 40 mL and stored at 4°C overnight. Dis- and reassembly of the protein containers was monitored by SEC, TEM, DLS and CD spectroscopy.

7.12 Encapsulation of plasmonic nanoparticles

The procedure was the same for AuNP as well as AgNP. Therefore, only the encapsulation of AuNP will be described in detail. 2 mg of the protein container was disassembled as described above. For reassembly, the desired amount of AuNP and sodium chloride was added to the reassembly solution. Protein container was formed around the nanoparticle during the incubation at 4°C overnight. Precipitated AuNP were isolated by centrifugation of the sample at 6000 G for 10 minutes. Subsequently, the sample was injected directly on a 5 mL *Sepharose HiTrap SP HP* cation exchange column for $AuFtn^{(pos)}$ and on a 5 mL *Sepharose HiTrap SP Q* anion exchange column for $AuFtn^{(neg)}$ to separate free AuNP from protein containers. The elution gradient was the same as for protein purification with a minor change for $Ftn^{(pos)}$. Here, the elution was started at 0.15 M NaCl. During the ion exchange chromatography, protein sample was detected at 280 nm and AuNPs at 520 nm. AuNP-filled protein container fractions were collected, and the protein solution was concentrated to 2 mL final volume. The sample was purified on a *Hiload 16/600 Superdex 200 PG* size exclusion column as for the protein purification. $AuFtn^{(neg)}$ obtained with the acidic disassembly condition was directly purified with size exclusion chromatography. For crystallization experiments, the sample was concentrated to a protein concentration of around 6 mg/mL. Samples were further analyzed by DLS and TEM.

7.13 Encapsulation of dyes

7.13.1 Rhodamine 6G in $Ftn^{(pos)}$

The encapsulation of rhodamine 6G in $Ftn^{(pos)}$ was achieved exploiting a thermal diffusion strategy. In detail, 2 mg $Ftn^{(pos)}$ were incubated in 2 mL of buffer J at 65°C for 30 minutes. 40 mg (0.084 mmol) rhodamine 6G were dissolved in 1 mL ethanol. 200 µL of the rhodamine solution mixed with 1.8 mL buffer J were added dropwise to the protein container solution. The mixture was stirred for 2.5 h. After the solution cooled down, it was purified by washing five times with H_2O MilliQ to remove free dye and concentrated to 2 mL. For further purification, the sample was injected on a *Hiload 16/600 Superdex 200 PG* size exclusion column. Protein sample was detected at 280 nm and the dye at 550 nm. Fractions with dye-encapsulated protein container were collected and concentrated to a protein concentration of 6 mg/mL for crystallization experiments.

7.13.2 Rhodamine B in Ftn$^{(neg)}$

The encapsulation of rhodamine B was performed similar to the AuNP encapsulation in acidic condition. Briefly, 5 mg Ftn$^{(neg)}$ was incubated in 1.5 mL buffer G at RT for 4 h to disassemble the protein container. For reassembly, 3 mL buffer H with rhodamine B (0.100 mmol) was added dropwise and the solution was stored at 4°C overnight. The purification was the same as for the dye encapsulation in Ftn$^{(pos)}$.

7.13.3 Determination of the dye loading

The amount of dye in both protein container variants was determined using UV-Vis spectroscopy. The dye concentration was determined from the absorption at λ_{max} (rhodamine 6G = 530 nm and rhodamine B = 550 nm) and the protein concentration from the absorption at 280 nm. Since the protein do not show any absorption above 350 nm, the dye concentration was directly determined by preparing a standard calibration curves for each dye in the range of 1 – 10 µM. The protein concentration was determined after subtracting the absorbance of the dye at 280 nm based on the assumption that the ratio of $A280/A\lambda_{max}$ is constant at any concentration. A standard calibration curve was prepared for the protein in the range of 0.1 – 1.0 mg/mL. Finally, the loading was determined by calculating the ratio of dye to protein container.

7.14 Catalysis with metal oxide nanoparticles in solution

7.14.1 Oxidase-like and peroxidase-like activity

The oxidase-like and peroxidase-like activity of the metal oxide loaded protein containers were investigated by monitoring the oxidation of 3,3',5,5'-tetramethylbenzidine (TMB). TMB is oxidized to give a blue colored product with absorbance maxima at 370 nm and 645 nm. A 200 µL solution containing 1 mM TMB and 0.04 mg/mL protein in citrate buffer of pH 4.0 was prepared and the reaction was monitored using a *JASCO V-630* spectrophotometer after an incubation time of one day. The peroxidase-like activity was measured by adding 50 mM H_2O_2 to the sample with an incubation time of 15 minutes.

Control reactions were done with cerium oxide nanoparticles purchased from *Sigma* (No. 796077). The experimental set-up was the same as for the metal oxide loaded protein containers.

7.14.2 Kinetic analysis

For the kinetic analysis, further experiments were performed using above-mentioned conditions by varying the concentrations of TMB (0.1 mM to 1 mM) and H_2O_2 (10 mM to 50 mM) while the concentration of the protein was kept constant (0.04 mg/mL). The kinetic parameters were calculated based on the Michaelis-Menten equation:

$$v = \frac{v_{max}[S]}{K_m + [S]} \quad (7\text{-}16)$$

where v is the initial velocity, v_{max} the maximal reaction velocity, $[S]$ the concentration of the substrate and K_m the Michaelis constant. K_m and v_{max} were obtained using the Lineweaver-Burk plot (Appendix, Figure 8.18).

7.15 Kinetic measurements of the AuNPs in solution

The catalytic activity of AuNPs was investigated for the nitroarene reduction. The procedure was the same for all reactions. AuNP, AuFtn$^{(pos)}$ and AuFtn$^{(neg)}$ stock solutions were adjusted to the same concentration using UV-Vis spectroscopy (Appendix, Figure 8.31). Four different substrates were used: nitrobenzyl (NB), 4-nitrophenol (4-NP), 3-nitrobenzenesulfonate (NBS) and (3-nitrobenzyl)trimethylammonium (NTA). As reducing agent $NaBH_4$ (10 mM) was used and the reactions were monitored by UV-Vis spectroscopy. The kinetic parameters were calculated based on a first-order reaction according to equation 7.17.:

$$\ln\left(\frac{A}{A_0}\right) = -kt \quad (7\text{-}17)$$

where A is the absorbance at the time t, A_0 is the absorbance at the beginning and k is the first-order rate constant. During the reaction, an intermediate was formed for NB, NBS and NTA. The reaction of the substrate to the intermediate is very fast, whereas the consecutive reaction to the product is comparably slow. Thus, the second reaction was investigated for the determination of the kinetic parameters. The absorbance at the beginning was determined by fitting an exponential decay to the absorption maximum of the intermediate (Appendix, Figure 8.35).

7.16 Protein Crystallography

7.16.1 Hanging drop vapor diffusion

Hanging drop vapor diffusion method for optimization of the crystallization condition was performed in pre-greased 4 x 6 well CrystalClear plates (*Jena Bioscience*). The purified protein solutions were concentrated to 6 mg/mL. The reservoir was filled with 500 µL of the crystallization cocktail. 2 µL of the crystallization cocktail was pipetted on the center of a siliconized glass cover slide (*Jena Bioscience*) and mixed with 1 µL of the positively charged protein container and 1 µL of the negatively surface charged protein container solution. For unitary crystals, the protein solution was replaced by the corresponding buffer. Then the cover slide was turned over and placed on the greased rim of the reservoir well. The drop was equilibrated against the reservoir solution at 20°C.

7.16.2 Batch crystallization

Batch crystallization was performed based on a method developed by Rayment.[268] The mother liquor is prepared in a 1.5 mL *Eppendorf* tube. To this end, 5 µL of $Ftn^{(pos)}$ solution (or 5 µL buffer for unitary crystals) are pipetted onto the bottom of the tube and 5 µL of $Ftn^{(neg)}$ solution are added while vortexed at moderate speed for ~10 s. 10 µL of the precipitant solution is added dropwise while vortexing, which is continued for another 5–10 s. The mixture is transferred to a crystallization well on the plate. The wells are sealed air-tight with transparent duct tape and the plate is stored at 20°C until observation.

7.16.3 Stabilization of crystals

Crystals were cross-linked *in situ* in the hanging drop of the crystallization setup. The cover slide with the crystals in the crystallization drop was briefly removed to add 10 µL of 25% glutaraldehyde to the reservoir solution (final concentration 0.5%), followed by mixing of the reservoir solution with the pipette tip. The cover slide with the crystallization drop was put back on the reservoir and the crystallization plate was incubated up to 48 h. Glutaraldehyde diffuses slowly into the crystallization drop and gently cross-links the crystals within the drop.

7.16.4 Catalytic activity

Metal oxide loaded crystals were stabilized and transferred to a drop of water several times to wash the crystals therein. Solutions for testing the catalytic activity were prepared as drops on siliconized glass cover slides containing 1 mM dye in citrate buffer of pH 4.0 (for peroxidase-like activity additionally 50 mM H_2O_2 were added). The following dyes were used: 3,3',5,5'-tetramethylbenzidine (TMB), *o*-phenyldiamine (OPD), *o*-dianisidine (DIA), 5-aminosalicylic acid (5-ASA), 3,3'-diaminobenzidin (DAB) and 3-amino-9-ethylcarbazole (3AEC). The crystals were transferred to the prepared drops and put over a reservoir with 500 µL water as reservoir solution. The coloration of the crystals, after few minutes up to one day, indicated conversion of the substrate to the colored product

AuNP-loaded crystals from one 4 x 6 well plate were stabilized and transferred with a pipette into a 1.5 mL *Eppendorf* tube. The crystals were washed several times with H_2O MilliQ and stored in 100 µL. Reaction solution was prepared separately and added to the *Eppendorf* tube to start the reaction. Supernatant was measured by UV-Vis spectroscopy to determine the turnover of the reaction.

8 Appendix

Table 8.1: List of mutations in Ftn$^{(pos)}$ and Ftn$^{(neg)}$.

Variant	Mutations
Ftn$^{(pos)}$	A18K, N25R, C90K, N98K, C102K, H105K, N109K, D123K, E162R
Ftn$^{(neg)}$	A18E, C90E, C102E, H105E

The sequence for human H chain ferritin with the H86K mutation was used as basis.

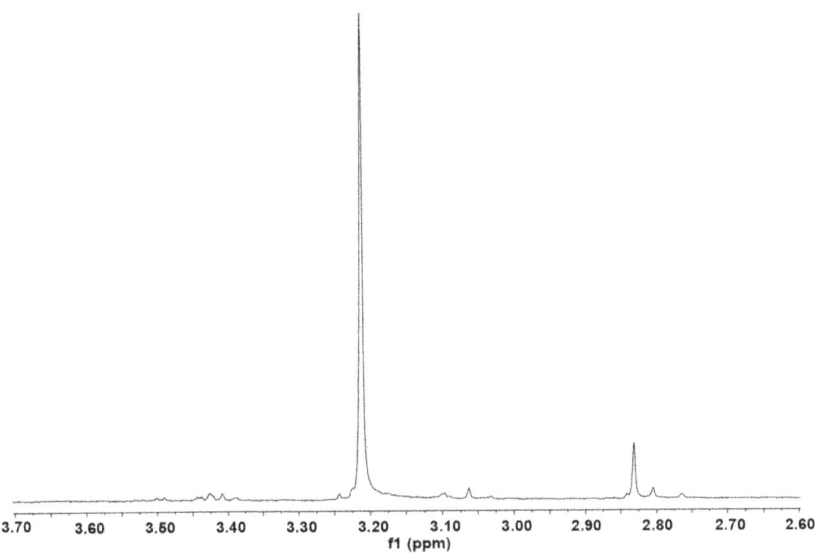

Figure 8.1: ^{1}H-NMR of AuC2mix treated with aqua regia. AuC2mix was treated with aqua regia to remove the gold core. Less signals than expected are observed and the signals do not show the expected chemical shift, indicating a side reaction.

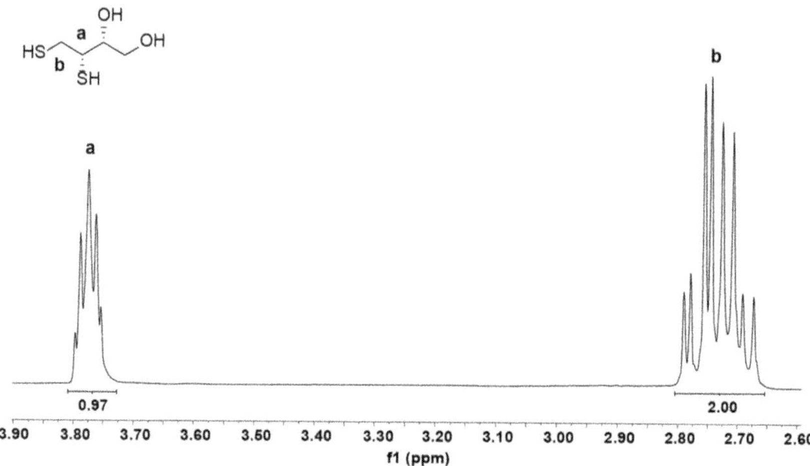

Figure 8.2: ¹H-NMR of DTT. Signals for the DTT molecule do not interfere with signals from the ligand.

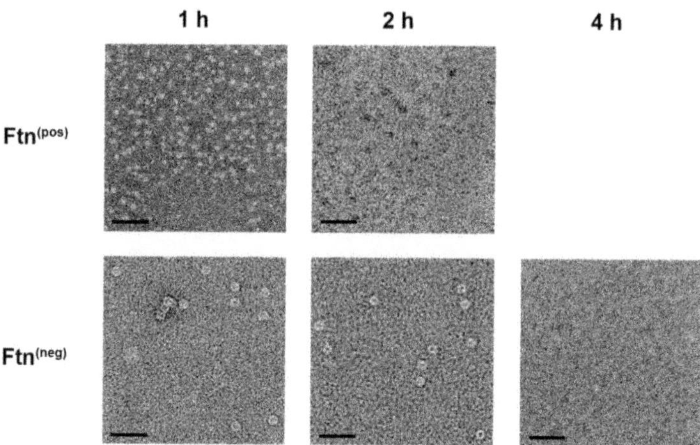

Figure 8.3: Disassembly of protein containers with different incubation time. TEM images of the disassembly of Ftn(pos) and Ftn(neg) with different incubation times. Scale bars are 50 nm.

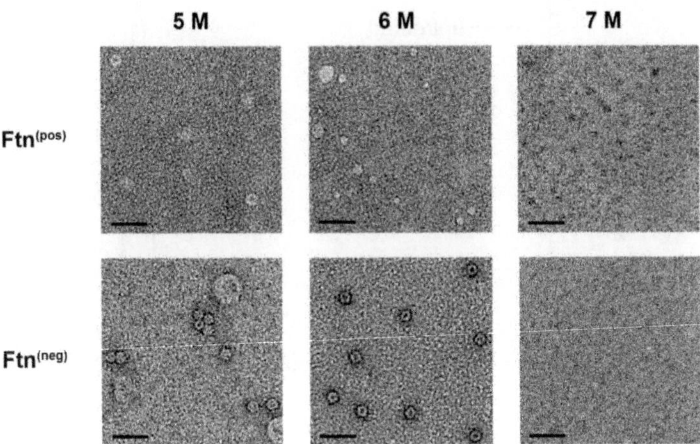

Figure 8.4: Disassembly of protein containers with different Gua concentrations. TEM images of the disassembly of Ftn(pos) and Ftn(neg) with different Gua concentrations. Scale bars are 50 nm.

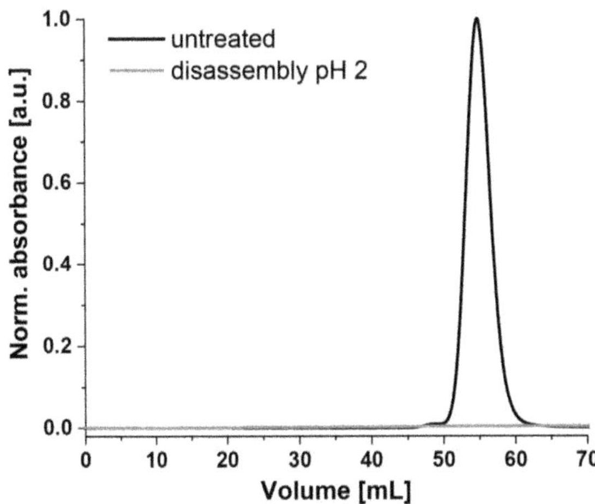

Figure 8.5: SEC of the disassembly of Ftn(pos) in pH 2. The untreated sample is displayed in black and the disassembly in pH 2 in red.

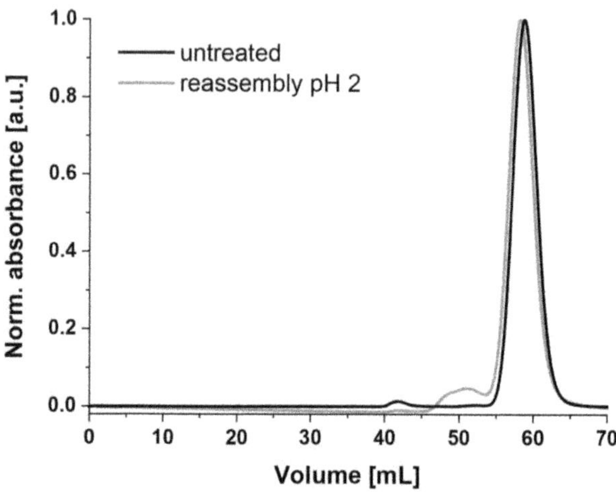

Figure 8.6: SEC of the reassembly of Ftn$^{(neg)}$ in pH 2. Further improvement of the reassembly procedure by adjusting the pH value to 7.4 with NaOH. The untreated sample is displayed in black and the reassembly in red.

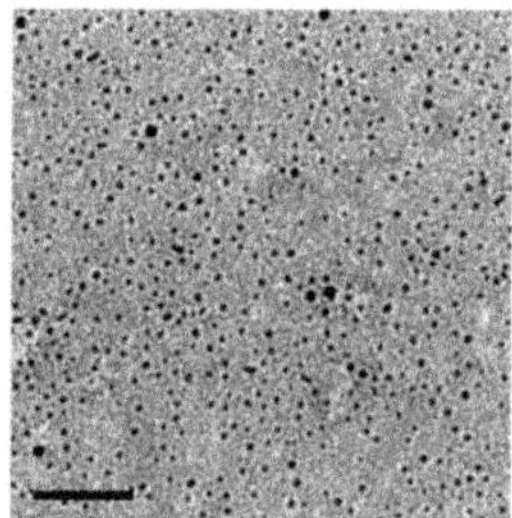

Figure 8.7: TEM image of the AuFtn$^{(pos)}$ SEC. TEM image of the peak at 95 mL in the SEC of AuFtn$^{(pos)}$. Scale bar is 50 nm.

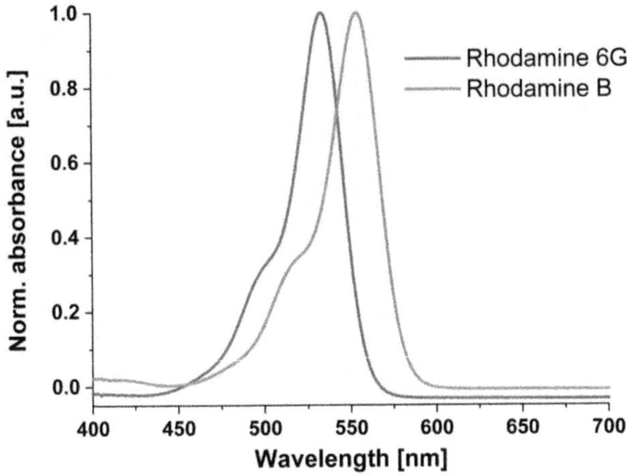

Figure 8.8: UV-Vis spectra of rhodamine 6G and rhodamine B. Rhodamine 6G is displayed in blue and rhodamine B in red.

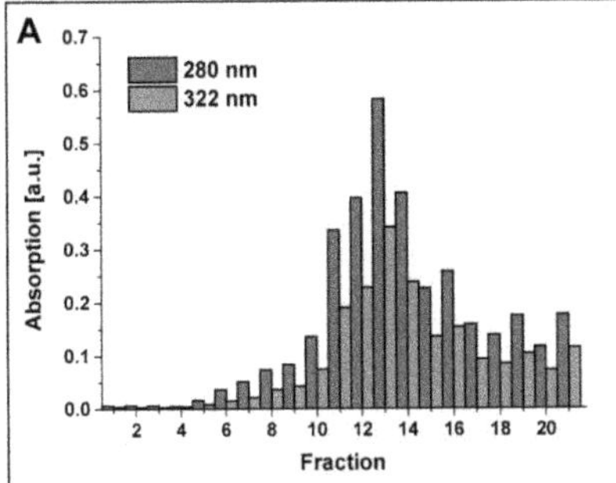

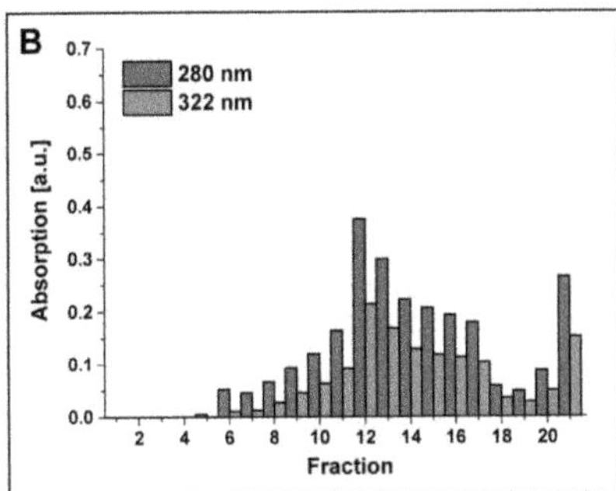

Figure 8.9: Sucrose gradient centrifugation of cerium oxide loaded protein containers. Analysis of sucrose gradient centrifugation of (A) CeFtn$^{(pos)}$ and (B) CeFtn$^{(neg)}$. UV-Vis absorption was measured at 280 nm and 322 nm.

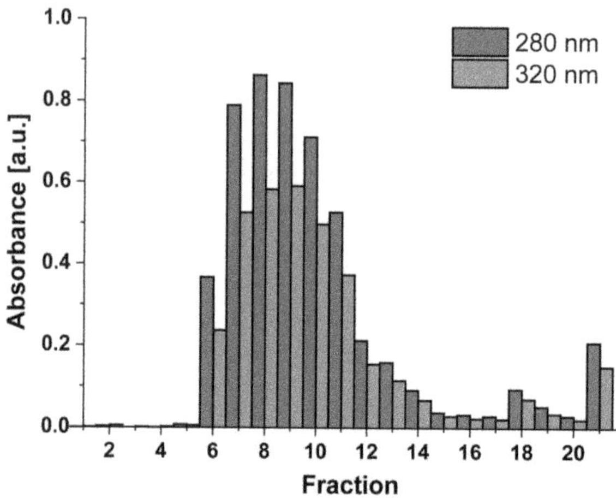

Figure 8.10: Sucrose gradient centrifugation of FeFtn$^{(neg)}$. UV-Vis absorption was measured at 280 nm and 322 nm.

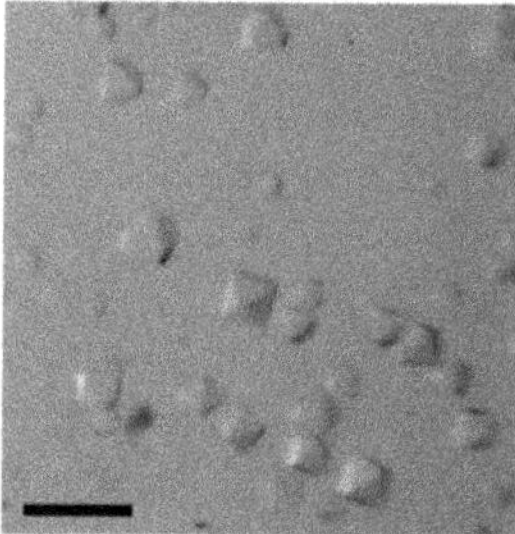

Figure 8.11: Batch crystallization of Ftn$^{(neg)}$ crystals. Batch crystallization of Ftn$^{(neg)}$ at 500 mM Mg^{2+} concentration (crystallization condition for hanging drop vapor diffusion method). Scale bar is 200 µm.

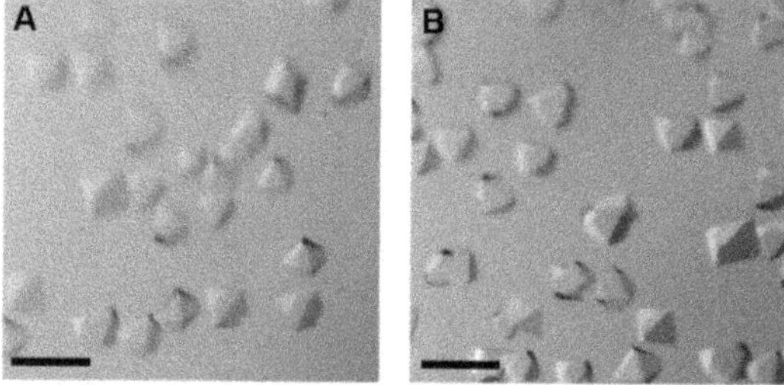

Figure 8.12: Batch crystallization of Ftn$^{(pos)}$/Ftn$^{(neg)}$ crystals. Batch crystallization of Ftn$^{(pos)}$ and Ftn$^{(neg)}$ at 180 mM Mg^{2+} concentration after (A) 1 day and (B) four weeks. Scale bars are 200 µm.

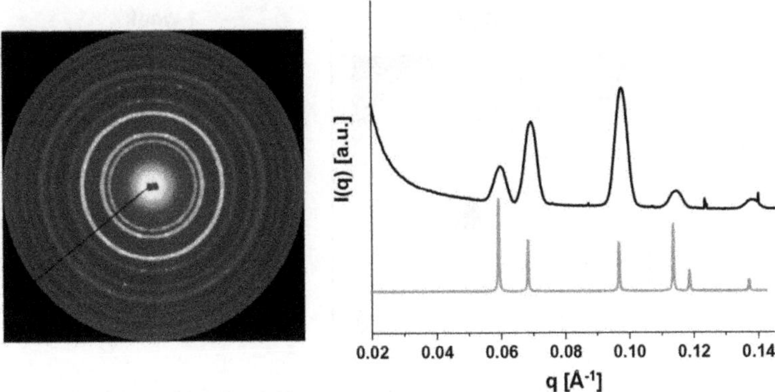

Figure 8.13: SAXS data of eFtn$^{(neg)}$. 2D SAXS pattern (left) and radially averaged 1D SAXS data (right) of eFtn$^{(neg)}$. The measured data are displayed in black and simulated data in red.

Table 8.2: Cell parameters obtained from SAXS data.

hkl	exp. *q* (Å^{-1})	calc. *q* (Å^{-1})	cell parameters
111	0.05988	0.06009	
200	0.06940		a = 181.1 Å
300	0.09757	0.10408	
311	0.11450	0.11507	
400	0.13810	0.13878	

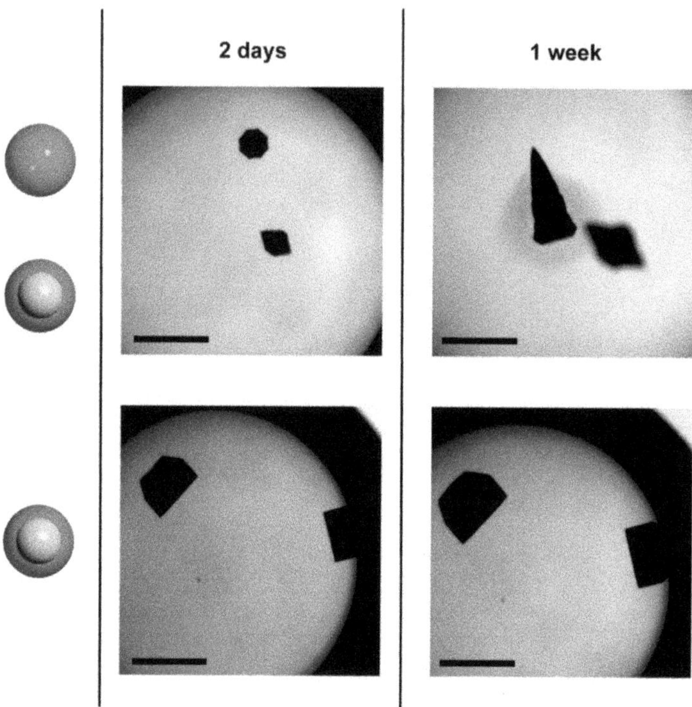

Figure 8.14: Crystallization of AuFtn$^{(neg)}$ obtained with the pH 2 method. eFtn$^{(pos)}$/AuFtn$^{(neg)}$ and AuFtn$^{(neg)}$ crystals after two days and one week. Scale bars are 200 µm.

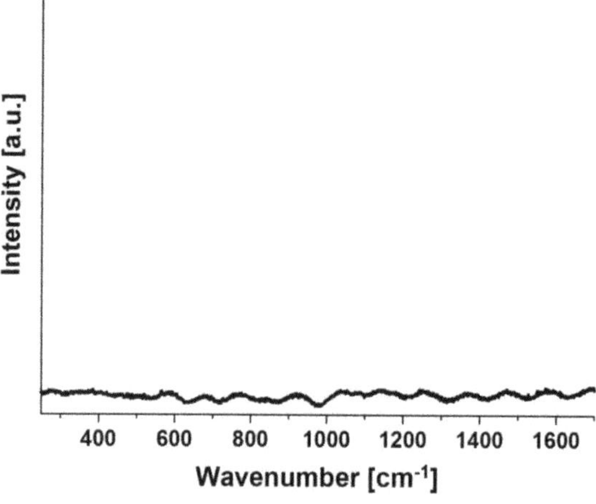

Figure 8.15: SERS measurement of AuFtn$^{(pos)}$/eFtn$^{(neg)}$ in water.

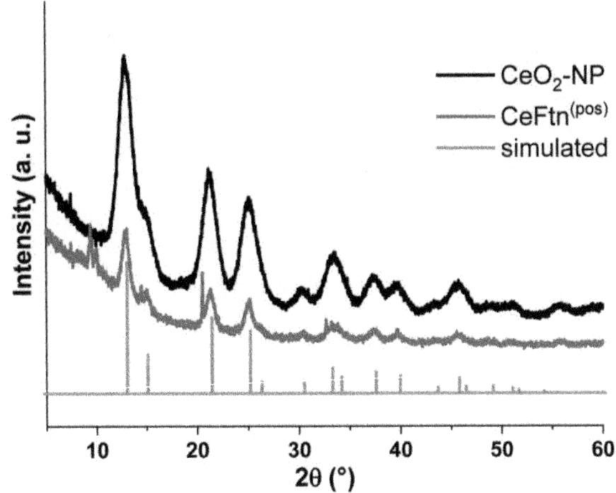

Figure 8.16: X-ray diffraction pattern of CeFtn$^{(pos)}$. Experimental data of commercially available nanoparticles are shown in black and of CeFtn$^{(pos)}$ in blue. Calculated diffraction pattern for CeO_2 is shown in red.

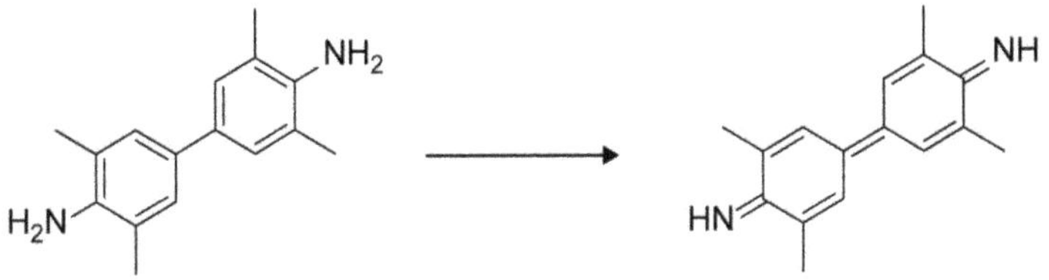

Figure 8.17: Oxidation of TMB. Oxidation of the colorless TMB to the blue colored product.

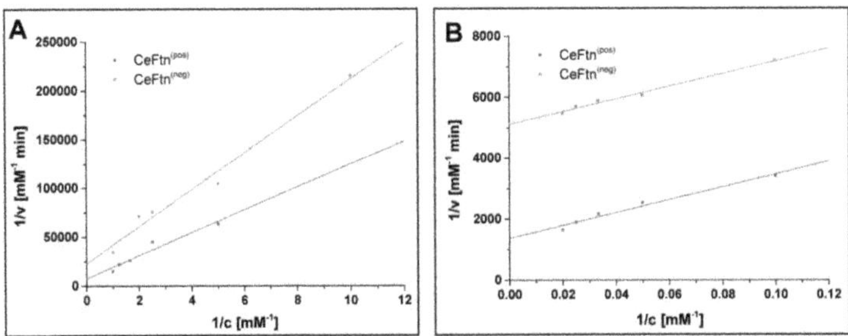

Figure 8.18: Determination of kinetic parameters for oxidase-like and peroxidase-like activity. Double reciprocal plot of (A) oxidase-like and (B) peroxidase-like activity.

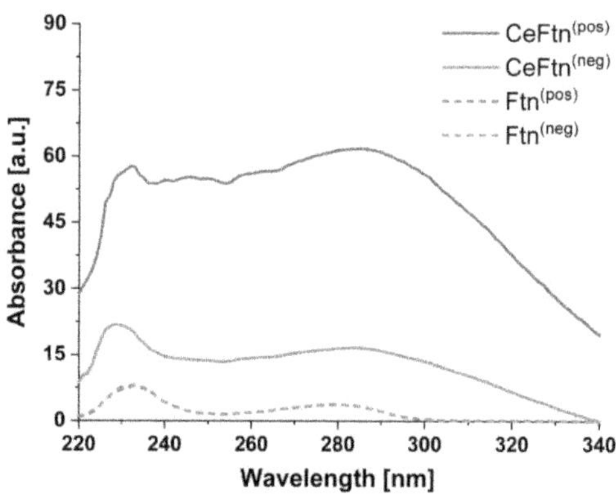

Figure 8.19: UV-Vis absorption spectra of empty and CeO_2 loaded protein containers. Ftn(pos) samples are displayed in blue and Ftn(neg) in red. All protein samples have a concentration of 4 mg/mL.

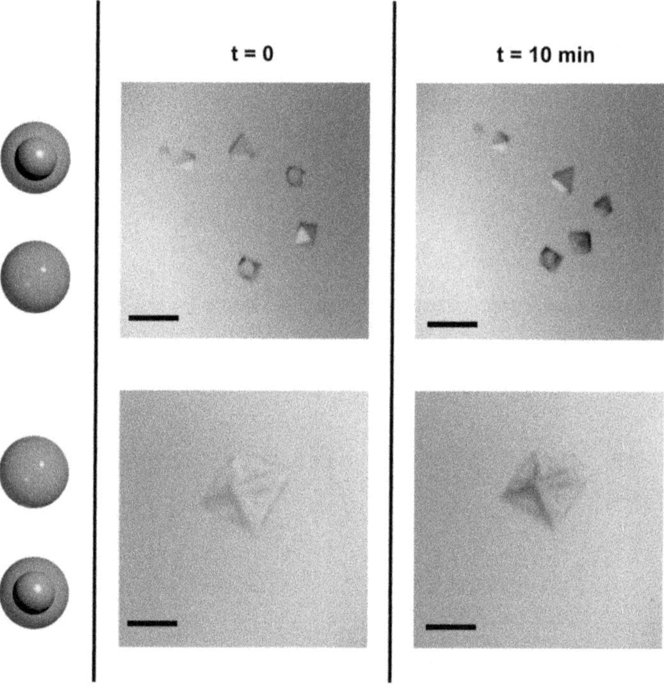

Figure 8.20: Microscopy images of oxidase-like activity of CeO_2 loaded protein crystals with TMB. Optical microscopy images of CeFt(pos)/eFtn(neg) (top) and eFtn(pos)/CeFtn(neg) (bottom) crystals in a drop with 1 mM TMB. Pictures were taken at the start of the reaction and after 10 minutes. Scale bars are 200 µm.

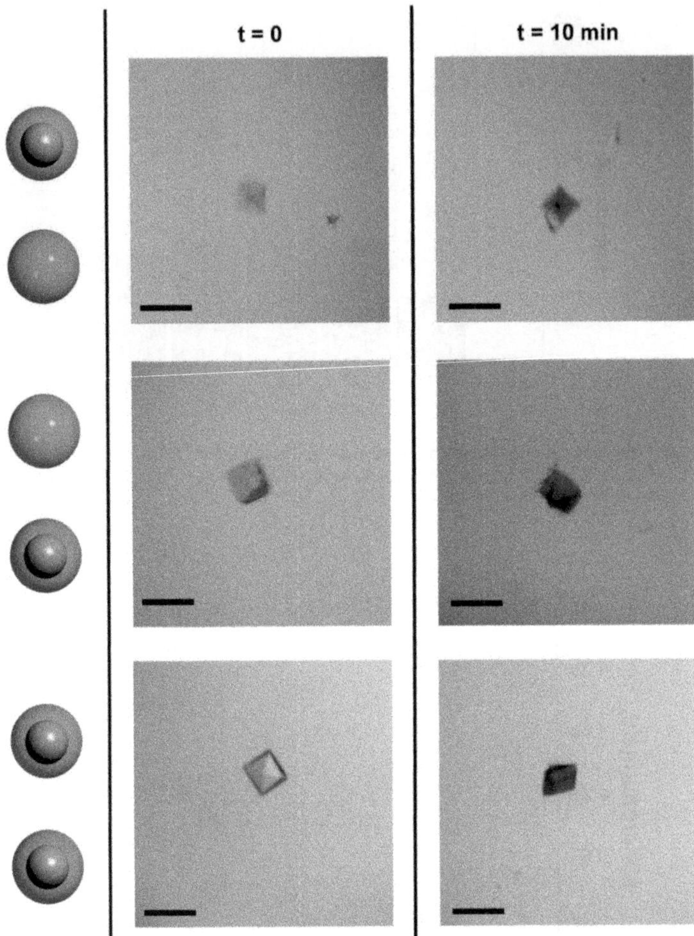

Figure 8.21: Microscopy images of peroxidase-like activity of CeO_2 loaded protein crystals with TMB. Optical microscopy images of CeFt$^{(pos)}$/eFtn$^{(neg)}$ (top), eFtn$^{(pos)}$/CeFtn$^{(neg)}$ (middle) and CeFtn$^{(pos)}$/CeFtn$^{(neg)}$ (bottom) crystals in a drop with 1 mM TMB and 50 mM H_2O_2. Pictures were taken at the start of the reaction and after 10 minutes. Scale bars are 200 µm.

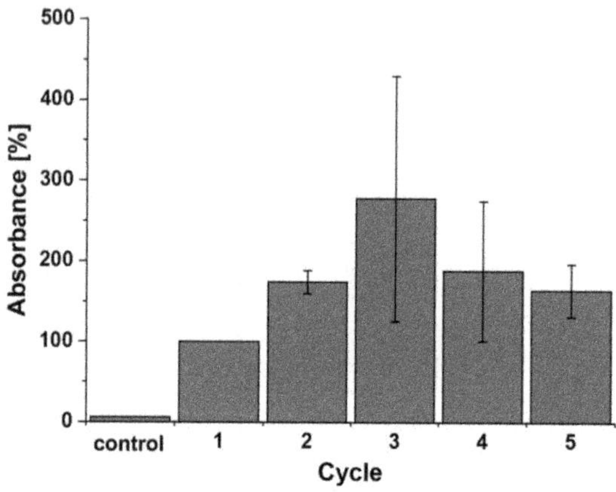

Figure 8.22: Catalytic cycling for oxidase-like activity. Assembly of CeFtn(pos)/CeFtn(neg) crystals was transferred several times into a 1 mM TMB solution. UV-Vis intensity at 645 nm was measured at normalized against the first measurement.

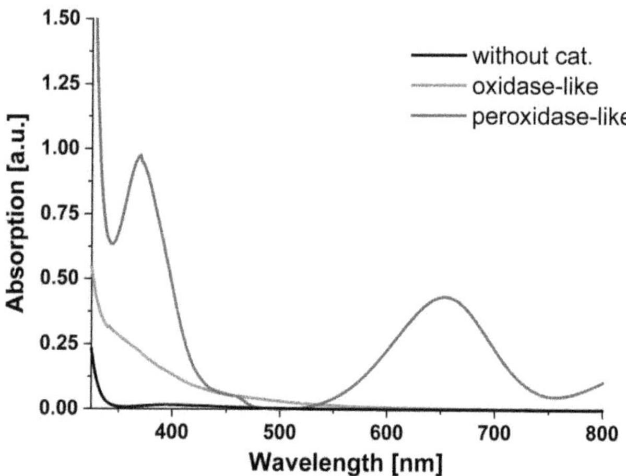

Figure 8.23: Oxidase-like and peroxidase-like activity of FeFtn(neg). UV-Vis absorption spectra of TMB was monitored. The reaction without catalyst is shown in black, oxidase-like activity in red and peroxidase-like activity in blue.

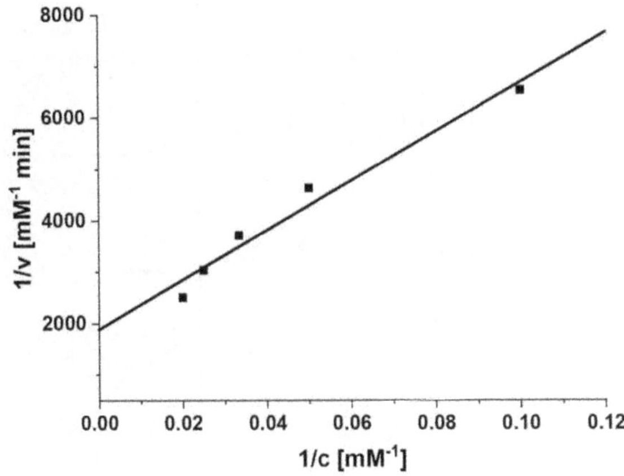

Figure 8.24: Steady state kinetic for the peroxidase-like activity of FeFtn$^{(neg)}$. Double reciprocal plot of peroxidase-like activity at constant concentration of TMB and FeFtn$^{(neg)}$ while the concentration of H_2O_2 was varied.

Table 8.3: Kinetic parameters of the peroxidase-like activity of FeFtn$^{(neg)}$.

Building block	Substrate	v_{max} [µM s^{-1}]	K_m [mM]
FeFtn$^{(neg)}$	H_2O_2 + TMB	8.8×10^{-3}	25.42

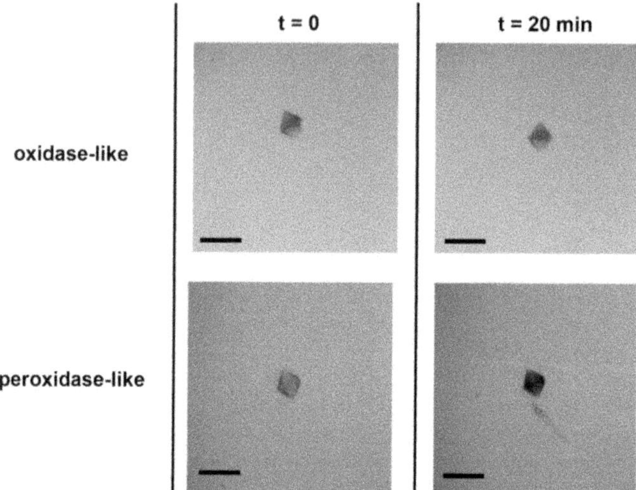

Figure 8.25: Microscopy images of oxidase-like and peroxidase-like activity of eFtn(pos)/FeFtn(neg) crystals. Pictures show a crystallization drop with 1 mM TMB without H_2O_2 for oxidase-like activity and with 50 mM H_2O_2 for peroxidase-like activity. Scale bars are 200 µm.

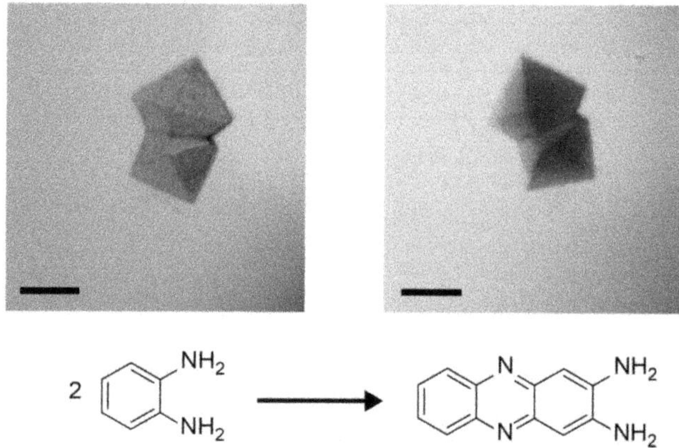

Figure 8.26: Oxidase activity of CeFtn(pos)/CeFtn(neg) crystals with OPD. Pictures show a crystallization drop with 1 mM OPD at 0 min (left) and after one day (right) of incubation. Scale bars are 200 µm.

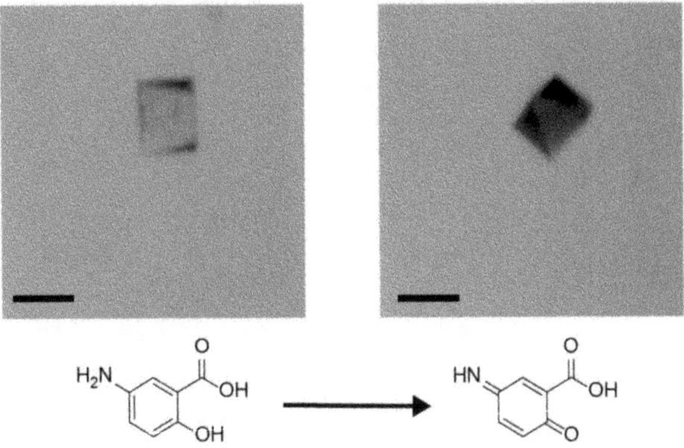

Figure 8.27: Oxidase activity of CeFtn(pos)/CeFtn(neg) crystals with 5-ASA. Pictures show a crystallization drop with 1 mM 5-ASA at 0 min (left) and after one day (right) of incubation. Scale bars are 100 µm.

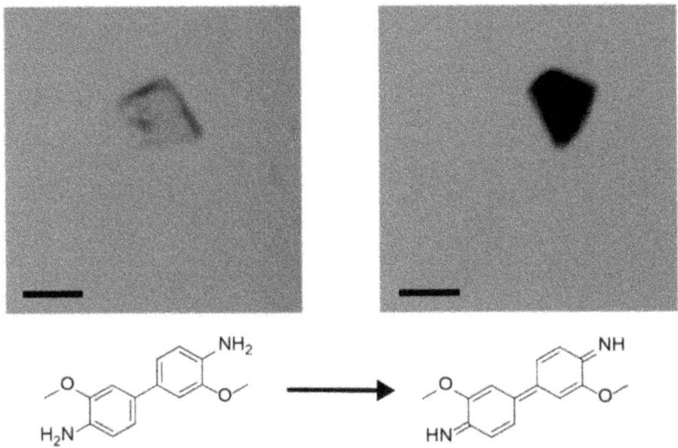

Figure 8.28: Oxidase activity of CeFtn(pos)/CeFtn(neg) crystals with DIA. Pictures show a crystallization drop with 1 mM DIA at 0 min (left) and after one day (right) of incubation. Scale bars are 100 µm.

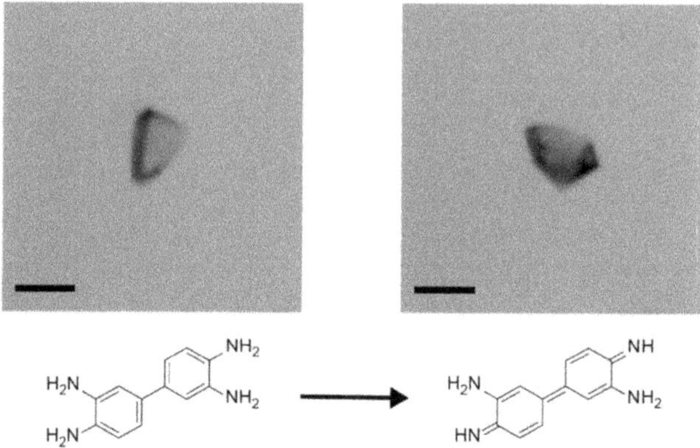

Figure 8.29: Oxidase activity of CeFtn$^{(pos)}$/CeFtn$^{(neg)}$ crystals with DAB. Pictures show a crystallization drop with 1 mM DAB at 0 min (left) and after 20 minutes (right) of incubation. Scale bars are 100 µm.

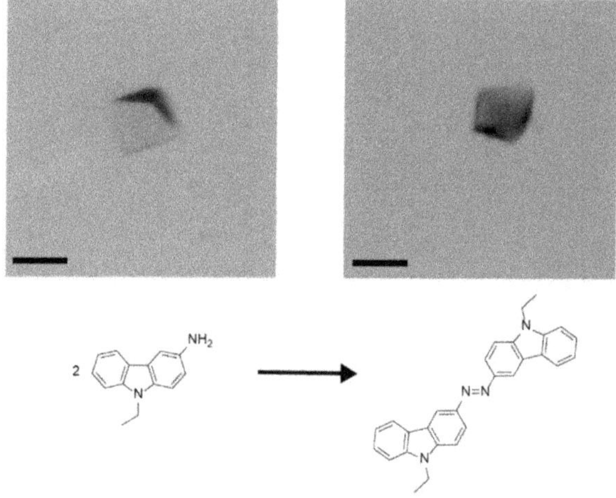

Figure 8.30: Oxidase activity of CeFtn$^{(pos)}$/CeFtn$^{(neg)}$ crystals with 3AEC. Pictures show a crystallization drop with 1 mM 3AEC at 0 min (left) and after 20 minutes (right) of incubation. Scale bars are 100 µm.

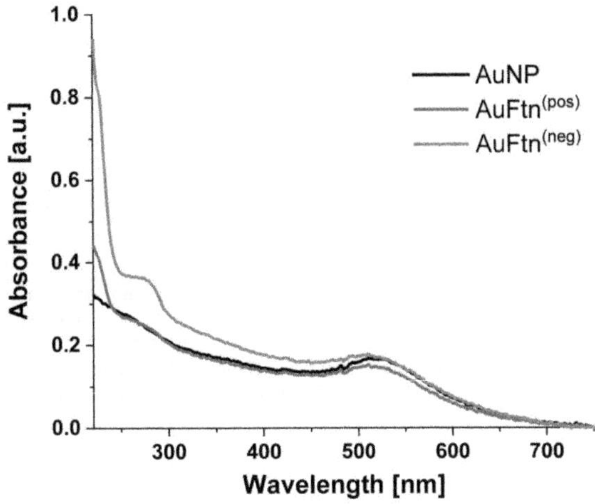

Figure 8.31: UV-Vis spectra of different AuNP samples. The concentration of AuNP (black), AuFtn(pos) (blue) and AuFtn(neg) (red) was adjusted to the same concentration by UV-Vis spectroscopy. Same intensity of the plasmon absorbance maximum indicates same concentration of AuNP.

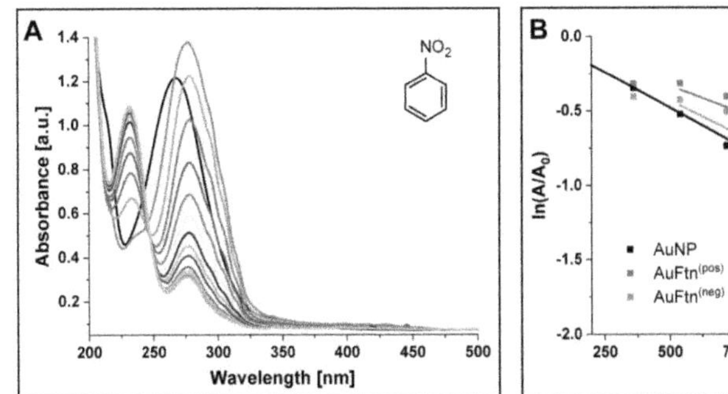

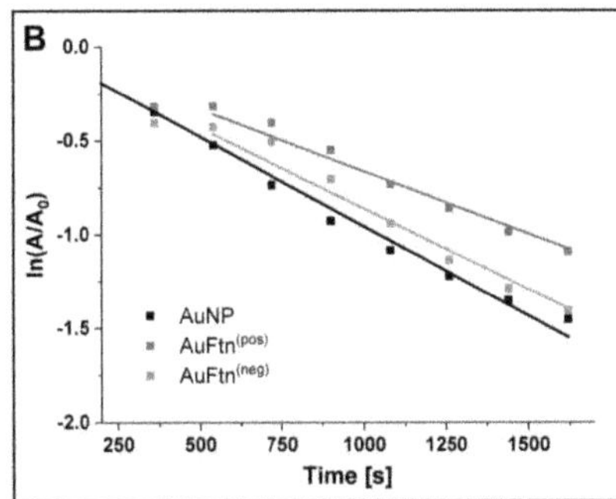

Figure 8.32: Determination of kinetic parameters for the reduction of NB. (A) UV-Vis spectra were measured for 45 minutes in 3 min intervals. (B) Reaction rate was determined for AuNP in black, AuFtn(pos) in blue and AuFtn(neg) in red.

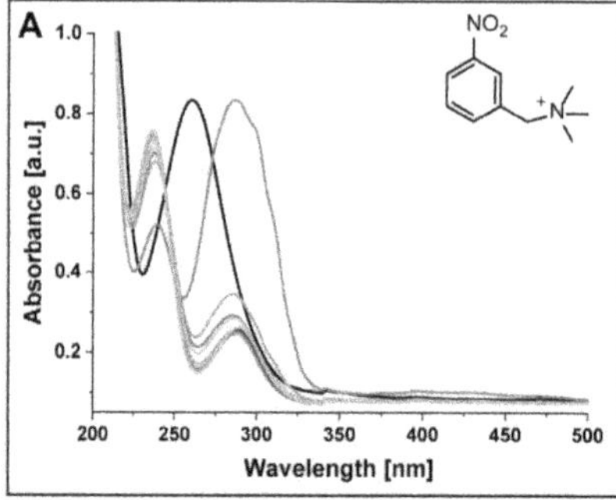

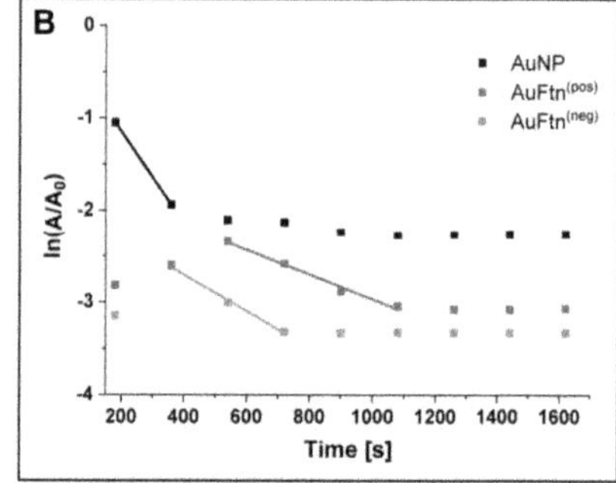

Figure 8.33: Determination of kinetic parameters for the reduction of NTA. (A) UV-Vis spectra were measured for 45 minutes in 3 min intervals. (B) Reaction rate was determined for AuNP in black, AuFtn(pos) in blue and AuFtn(neg) in red.

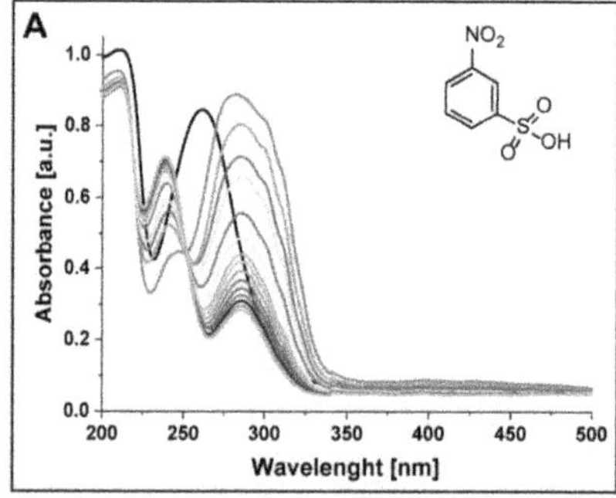

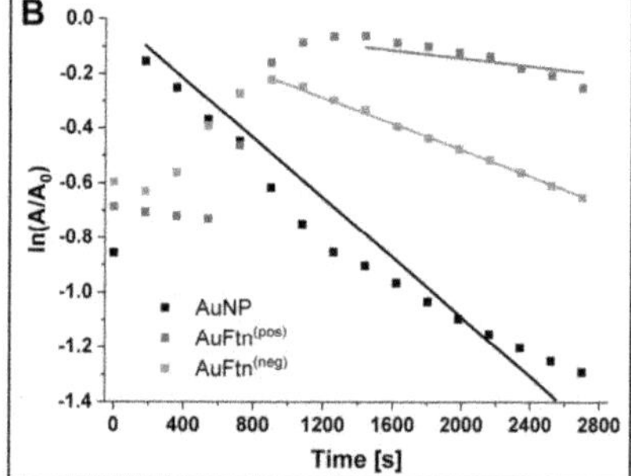

Figure 8.34: Determination of kinetic parameters for the reduction of NBS.A) UV-Vis spectra were measured for 45 minutes in 3 min intervals. (B) Reaction rate was determined for AuNP in black, AuFtn(pos) in blue and AuFtn(neg) in red.

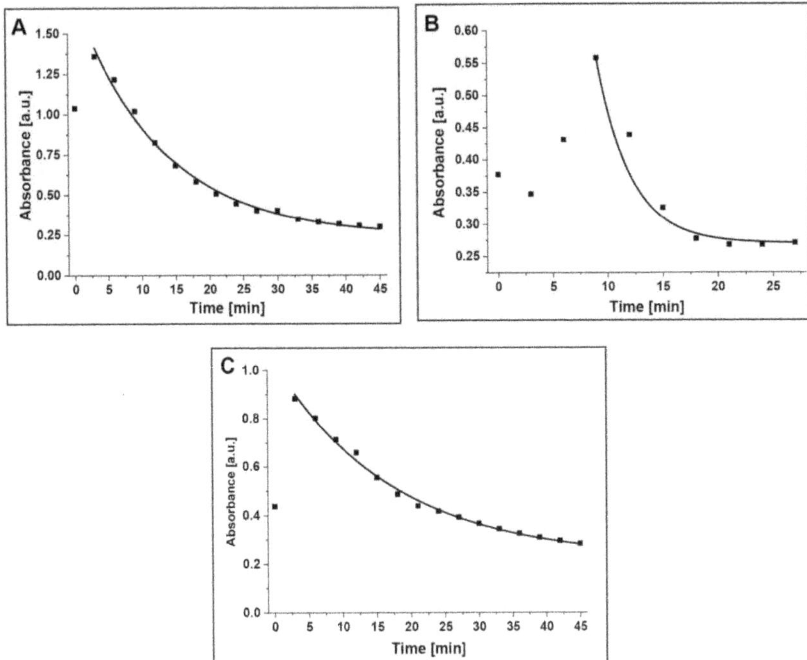

Figure 8.35: Determination of the starting value. The starting value of the intermediate for the kinetic analysis was determined by an exponential plot for (A) NB, (B) NTA and (C) NBS.

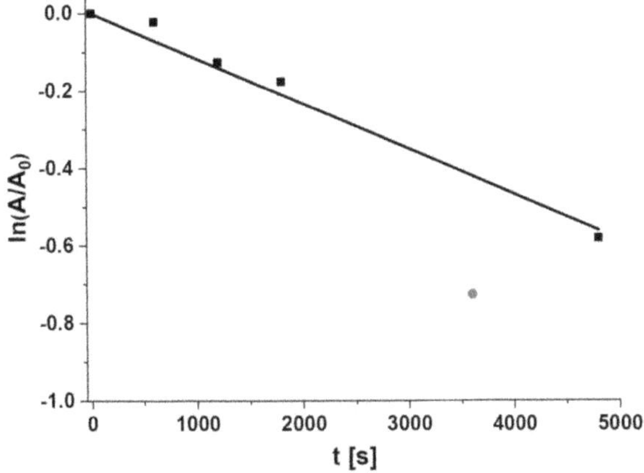

Figure 8.36: Determination of the kinetic parameters. Determination of the reaction rate using AuFtn(pos)/eFtn(neg) crystals as catalyst. The red data point is an outliner and was excluded for the linear regression.

9 Bibliography

[1] Talapin, D. V.; Lee, J. S.; Kovalenko, M. V.; Shevchenko, E. V. Prospects of Colloidal Nanocrystals for Electronic and Optoelectronic Applications. *Chem. Rev.* **2010**, *110*, 389–458.

[2] Lu, A. H.; Schmidt, W.; Matoussevitch, N.; Bönnemann, H.; Spliethoff, B.; Tesche, B.; Bill, E.; Kiefer, W.; Schüth, F. Nanoengineering of a Magnetically Separable Hydrogenation Catalyst. *Angew. Chem. Int. Ed.* **2004**, *43*, 4303–4306.

[3] Li, K.; Zhang, Z.-P.; Luo, M.; Yu, X.; Han, Y.; Wei, H.-P.; Cui, Z.-Q.; Zhang, X.-E. Multifunctional ferritin cage nanostructures for fluorescence and MR imaging of tumor cells. *Nanoscale* **2012**, *4*, 188–193.

[4] Murthy, S. K. Nanoparticles in modern medicine: state of the art and future challenges. *Int. J. Nanomed.* **2007**, *2*, 129–141.

[5] Puntes, V. F.; Gorostiza, P.; Aruguete, D. M.; Bastus, N. G.; Alivisatos, A. P. Collective behaviour in two-dimensional cobalt nanoparticle assemblies observed by magnetic force microscopy. *Nat. Mater.* **2004**, *3*, 263–268.

[6] Gupta, A. K.; Gupta, M. Synthesis and surface engineering of iron oxide nanoparticles for biomedical applications. *Biomaterials* **2005**, vol. 26 3995–4021.

[7] Shafi, K. V. P. M.; Gedanken, A.; Prozorov, R. Surfactant-Assisted Self-Organization of Cobalt Nanoparticles in a Magnetic Fluid. *Adv. Mater.* **1998**, *10*, 590–593.

[8] Li, M.; Mann, S. Emergent Hybrid Nanostructures Based on Non-Equilibrium Block Copolymer Self-Assembly. *Angew. Chem. Int. Ed.* **2008**, *47*, 9476–9479.

[9] Castro, C.; Ramos, J.; Millán, A.; González-Calbet, J.; Palacio, F. Production of Magnetic Nanoparticles in Imine Polymer Matrixes. *Chem. Mater.* **2000**, *12*, 3681–3688.

[10] Schreiber, R.; Do, J.; Roller, E.-M.; Zhang, T.; Schüller, V. J.; Nickels, P. C.; Feldmann, J.; Liedl, T. Hierarchical assembly of metal nanoparticles, quantum dots and organic dyes using DNA origami scaffolds. *Nat. Nanotechnol.* **2014**, *9*, 74–78.

[11] Talapin, D. V. Lego Materials. *ACS Nano* **2008**, *2*, 1097–1100.

[12] Zhang, S. Fabrication of novel biomaterials through molecular self-assembly. *Nat. Biotechnol.* **2003**, *21*, 1171–1178.

[13] Shevchenko, E. V; Talapin, D. V; Kotov, N. A.; O'Brien, S.; Murray, C. B. Structural diversity in binary nanoparticle superlattices. *Nature* **2006**, *439*, 55–59.

[14] Lehn, J. M. Supramolecular Chemistry. *Science* **1993**, *260*, 1762–1763.

[15] Koumura, N.; Zijistra, R. W. J.; Van Delden, R. A.; Harada, N.; Feringa, B. L. Light-driven monodirectional molecular rotor. *Nature* **1999**, *401*, 152–155.

[16] Fletcher, S. P.; Dumur, F.; Pollard, M. M.; Feringa, B. L. A Reversible, Unidirectional Molecular Rotary Motor Driven by Chemical Energy. *Science* **2005**, *310*, 80–82.

[17] Tseng, R. J.; Tsai, C.; Ma, L.; Ouyang, J.; Ozkan, C. S.; Yang, Y. Digital memory device based on tobacco mosaic virus conjugated with nanoparticles. *Nat. Nanotechnol.* **2006**, *1*, 72–77.

[18] Miao, L.; Han, J.; Zhang, H.; Zhao, L.; Si, C.; Zhang, X.; Hou, C.; Luo, Q.; Xu, J.; Liu, J. Quantum-Dot-Induced Self-Assembly of Cricoid Protein for Light Harvesting. *ACS Nano* **2014**, *8*, 3743-3751.

[19] Myers, B. D.; Lin, Q.-Y.; Wu, H.; Luijten, E.; Mirkin, C. A.; Dravid, V. P. Size-Selective Nanoparticle Assembly on Substrates by DNA Density Patterning. *ACS Nano* **2016**, *10*, 5679-5686.

[20] Nykypanchuk, D.; Maye, M. M.; van der Lelie, D.; Gang, O. DNA-guided crystallization of colloidal nanoparticles. *Nature* **2008**, *451*, 549–552.

[21] Park, S. Y.; Lytton-Jean, A. K.; Lee, B.; Weigand, S.; Schatz, G. C.; Mirkin, C. A. DNA-programmable nanoparticle crystallization. *Nature* **2008**, *451*, 553–556.

[22] Tian, Y.; Zhang, Y.; Wang, T.; Xin, H. L.; Li, H.; Gang, O. Lattice engineering through nanoparticle–DNA frameworks. *Nat. Mater.* **2016**, *15*, 654–662.

[23] Macfarlane, R. J.; Jones, M. R.; Lee, B.; Auyeung, E.; Mirkin, C. A. Topotactic Interconversion of Nanoparticle Superlattices. *Science* **2013**, *341*, 1222–1225.

[24] Lach, M.; Künzle, M.; Beck, T. Proteins as Sustainable Building Blocks for the Next Generation of Bioinorganic Nanomaterials. *Biochemistry* **2018**, *58*, 140–141.

[25] Jutz, G.; van Rijn, P.; Santos Miranda, B.; Böker, A. Ferritin: A Versatile Building Block for Bionanotechnology. *Chem. Rev.* **2015**, *115*, 1653–1701.

[26] Uchida, M.; Kang, S.; Reichhardt, C.; Harlen, K.; Douglas, T. The ferritin superfamily: Supramolecular templates for materials synthesis. *Biochim. Biophys. Acta* **2010**, *1800*, 834–845.

[27] Kramer, R. M.; Li, C.; Carter, D. C.; Stone, M. O.; Naik, R. R. Engineered Protein Cages for Nanomaterial Synthesis. *J. Am. Chem. Soc.* **2004**, *126*, 13282–13286.

[28] Galvez, N.; Sanchez, P.; Dominguez-Vera, J. M.; Soriano-Portillo, A.; Clemente-Leon, M.; Coronado, E. Apoferritin-encapsulated Ni and Co superparamagnetic nanoparticles. *J. Mater. Chem.* **2006**, *16*, 2757–2761.

[29] Butts, C. A.; Swift, J.; Kang, S.-G.; Costanzo, L. Di; Christianson, D. W.; Saven, J. G.; Dmochowski, I. J. Directing Noble Metal Ion Chemistry within a Designed Ferritin Protein. *Biochemistry* **2008**, *47*, 12729–12739.

[30] Kang, Y. J.; Uchida, M.; Shin, H.-H.; Douglas, T.; Kang, S. Biomimetic FePt nanoparticle synthesis within Pyrococcus furiosus ferritins and their layer-by-layer formation. *Soft Matter* **2011**, *7*, 11078–11081.

[31] Kasyutich, O.; Ilari, A.; Fiohllo, A.; Tatchev, D.; Hoell, A.; Ceci, P. Silver Ion Incorporation and Nanoparticle Formation inside the Cavity of Pyrococcus furiosus Ferritin: Structural and Size-Distribution Analyses. *J. Am. Chem. Soc.* **2010**, *132*, 3621–3627.

[32] Uchida, M.; Flenniken, M. L.; Allen, M.; Willits, D. A.; Crowley, B. E.; Brumfield, S.; Willis, A. F.; Jackiw, L.; Jutila, M.; Young, M. J.; Douglas, T. Targeting of Cancer Cells with Ferrimagnetic Ferritin Cage Nanoparticles. *J. Am. Chem. Soc.* **2006**, *128*, 16626–33.

[33] Douglas, T.; Stark, V. T. Nanophase Cobalt Oxyhydroxide Mineral Synthesized within the Protein Cage of Ferritin. *Inorg. Chem.* **2000**, *39*, 1828–1830.

[34] Meldrum, F. C.; Wade, V. J.; Nimmo, D. L.; Heywood, B. R.; Mann, S. Synthesis of inorganic nanophase materials in supramolecular protein cages. *Nature* **1991**, *349*, 684–687.

[35] Iwahori, K.; Yoshizawa, K.; Muraoka, M.; Yamashita, I. Fabrication of ZnSe Nanoparticles in the Apoferritin Cavity by Designing a Slow Chemical Reaction System. *Inorg. Chem.* **2005**, *44*, 6393–6400.

[36] Turyanska, L.; Bradshaw, T. D.; Sharpe, J.; Li, M.; Mann, S.; Thomas, N. R.; Patanè, A. The Biocompatibility of Apoferritin-Encapsulated PbS Quantum Dots. *Small* **2009**, *5*, 1738–1741.

[37] Iwahori, K.; Enomoto, T.; Furusho, H.; Miura, A.; Nishio, K.; Mishima, Y.; Yamashita, I. Cadmium Sulfide Nanoparticle Synthesis in Dps Protein from Listeria innocua. *Chem. Mater.* **2007**, *19*, 3105–3111.

[38] Wong, K. K. W.; Mann, S. Biomimetic Synthesis of Cadimium Sulfide-Ferritin Nanocomposites. *Adv. Mater.* **1996**, *8*, 928–932.

[39] Künzle, M.; Beck, T.; Lach, M. Crystalline protein scaffolds as a defined environment for the synthesis of bioinorganic materials. *Dalt. Trans.* **2018**, *47*, 10382–10387.

[40] Künzle, M.; Eckert, T.; Beck, T. Binary Protein Crystals for the Assembly of Inorganic Nanoparticle Superlattices. *J. Am. Chem. Soc.* **2016**, *138*, 12731–12734.

[41] Goesmann, H.; Feldmann, C. Nanopartikuläre Funktionsmaterialien. *Angew. Chem.* **2010**, *122*, 1402–1437.

[42] Comotti, M.; Della Pina, C.; Matarrese, R.; Rossi, M. The Catalytic Activity of 'Naked' Gold Particles. *Angew. Chem. Int. Ed.* **2004**, *43*, 5812–5815.

[43] Wu, Z.; Lee, D.; Rubner, M. F.; Cohen, R. E. Structural Color in Porous, Superhydrophilic, and Self-cleaning SiO2/TiO2 Bragg Stacks. *Small* **2007**, *3*, 1445–1451.

[44] Yu, W. W.; Falkner, J. C.; Yavuz, C. T.; Colvin, V. L. Synthesis of monodisperse iron oxide nanocrystals by thermal decomposition of iron carboxylate salts. *Chem. Commun.* **2004**, *10*, 2306–2307.

[45] Piella, J.; Bastús, N. G.; Puntes, V. Size-Controlled Synthesis of Sub-10-nanometer Citrate-Stabilized Gold Nanoparticles and Related Optical Properties. *Chem. Mater.* **2016**, *28*, 1066-1075.

[46] Adegoke, O.; Takemura, K.; Park, E. Y. Plasmonic Oleylamine-Capped Gold and Silver Nanoparticle-Assisted Synthesis of Luminescent Alloyed CdZnSeS Quantum Dots. *ACS Omega* **2018**, *3*, 1357–1366.

[47] Smith, D. R.; Pendry, J. B.; Wiltshire, M. C. K. Metamaterials and Negative Refractive Index. *Science* **2004**, *305*, 788–792.

[48] Atwater, H. A.; Polman, A. Plasmonics for improved photovoltaic devices. *Nat. Mater.* **2010**, *9*, 205–213.

[49] Alvarez-Puebla, R. A.; Zubarev, E. R.; Kotov, N. A.; Liz-Marzán, L. M. Self-assembled nanorod supercrystals for ultrasensitive SERS diagnostics. *Nano Today* **2012**, *7*, 6–9.

[50] Trovarelli, A.; Zamar, F.; Llorca, J.; Leitenburg, C. de; Dolcetti, G.; Kiss, J. T. Nanophase Fluorite-Structured CeO2–ZrO2Catalysts Prepared by High-Energy Mechanical Milling. *J. Catal.* **1997**, *169*, 490–502.

[51] Petersen, S.; Barcikowski, S. In Situ Bioconjugation: Single Step Approach to Tailored Nanoparticle-Bioconjugates by Ultrashort Pulsed Laser Ablation. *Adv. Funct. Mater.* **2009**, *19*, 1167–1172.

[52] Leng, W.; Pati, P.; Vikesland, P. J. Room temperature seed mediated growth of gold nanoparticles: mechanistic investigations and life cycle assesment. *Environ. Sci.: Nano* **2015**, *2*, 440–453.

[53] Jana, N. R.; Peng, X. Single-Phase and Gram-Scale Routes toward Nearly Monodisperse Au and Other Noble Metal Nanocrystals. *J. Am. Chem. Soc.* **2003**, *125*, 14280–14281.

[54] Harris, N.; Ford, M. J.; Cortie, M. B.; McDonagh, A. M. Laser-induced assembly of gold nanoparticles into colloidal crystals. *Nanotechnology* **2007**, *18*, 365301.

[55] Kobayashi, Y.; Horie, M.; Konno, M.; Rodríguez-González, B.; Liz-Marzán, L. M. Preparation and Properties of Silica-Coated Cobalt Nanoparticles. *J. Phys. Chem. B* **2003**, *107*, 7420–7425.

[56] Wiley, B.; Sun, Y.; Mayers, B.; Xia, Y. Shape-Controlled Synthesis of Metal Nanostructures: The Case of Silver. *Chem. - Eur. J.* **2005**, *11*, 454–463.

[57] Jin, R.; Cao, Y.; Mirkin, C. A.; Kelly, K. L.; Schatz, G. C.; Zheng, J. G. Photoinduced Conversion of Silver Nanospheres to Nanoprisms. *Science* **2001**, *294*, 1901–1903.

[58] Lee, S. M.; Jun, Y. W.; Cho, S. N.; Cheon, J. Single-Crystalline Star-Shaped Nanocrystals and Their Evolution: Programming the Geometry of Nano-Building Blocks. *J. Am. Chem. Soc.* **2002**, *124*, 11244–11245.

[59] Sau, T. K.; Murphy, C. J. Room Temperature, High-Yield Synthesis of Multiple Shapes of Gold Nanoparticles in Aqueous Solution. *J. Am. Chem. Soc.* **2004**, *126*, 8648–8649.

[60] Grzelczak, M.; Pérez-Juste, J.; Mulvaney, P.; Liz-Marzán, L. M. Shape control in gold nanoparticle synthesis. *Chem. Soc. Rev.* **2008**, *37*, 1783–1791.

[61] Turkevich, J.; Stevenson, P. C.; Hillier, J. A Study of the Nucleation and Growth Processes in the Synthesis of Colloidal Gold. *Discuss. Faraday Soc.* **1951**, *11*, 55–75.

[62] Daniel, M. C. M.; Astruc, D. Gold Nanoparticles: Assembly, Supramolecular Chemistry, Quantum-Size Related Properties and Applications toward Biology, Catalysis and Nanotechnology. *Chem. Rev.* **2004**, *104*, 293–346.

[63] Syafiuddin, A.; Salmiati; Salim, M. R.; Beng Hong Kueh, A.; Hadibarata, T.; Nur, H. A Review of Silver Nanoparticles: Research Trends, Global Consumption, Synthesis, Properties, and Future Challenges. *J. Chin. Chem. Soc.* **2017**, *64*, 732–756.

[64] Thanh, N. T. K.; Maclean, N.; Mahiddine, S. Mechanisms of Nucleation and Growth of Nanoparticles in solution. *Chem. Rev.* **2014**, *114*, 7610–7630.

[65] LaMer, V. K.; Dinegar, R. H. Theory, Production and Mechanism of Formation of Monodispersed Hydrosols. *J. Am. Chem. Soc.* **1950**, *72*, 4847–4854.

[66] Polte, J. Fundamental growth principles of colloidal metal nanoparticles - a new perspective. *CrystEngComm* **2015**, *17*, 6809–6830.

[67] Gentry, S. T.; Kendra, S. F.; Bezpalko, M. W. Ostwald Ripening in Metallic Nanoparticles: Stochastic Kinetics. *J. Phys. Chem. C* **2011**, *115*, 12736–12741.

[68] Li, D.; Nielsen, M. H.; Lee, J. R. I.; Frandsen, C.; Banfield, J. F.; De Yoreo, J. J. Direction-Specific Interactions Control Crystal Growth by Oriented Attachment. *Science* **2012**, *336*, 1014–1018.

[69] Perala, S. R. K.; Kumar, S. On the Mechanism of Metal Nanoparticle Synthesis in the Brust-Schiffrin Method. *Langmuir* **2013**, *29*, 9863–9873.

[70] Sperling, R. A.; Parak, W. J. Surface modification, functionalization and bioconjugation of colloidal inorganic nanoparticles. *Philos. Trans. R. Soc. A Math. Phys. Eng. Sci.* **2010**, *368*, 1333–1383.

[71] Yang, G.; Chang, W. S.; Hallinan, D. T. A convenient phase transfer protocol to functionalize gold nanoparticles with short alkylamine ligands. *J. Colloid Interf. Sci.* **2015**, *460*, 164–172.

[72] Manson, J.; Kumar, D.; Meenan, B. J.; Dixon, D. Polyethylene glycol functionalized gold nanoparticles: The influence of capping density on stability in various media. *Gold Bull* **2011**, *44*, 99–105.

[73] Fan, K.; Cao, C.; Pan, Y.; Lu, D.; Yang, D.; Feng, J.; Song, L.; Liang, M.; Yan, X. Magnetoferritin nanoparticles for targeting and visualizing tumour tissues. *Nat. Nanotechnol.* **2012**, *7*, 459–464.

[74] Laaksonen, T.; Ahonen, P.; Johans, C.; Kontturi, K. Stability and Electrostatics of Mercaptoundecanoic Acid-Capped Gold Nanoparticles with Varying Counterion Size. *ChemPhysChem* **2006**, *7*, 2143–2149.

[75] Brust, M.; Walker, M.; Bethell, D.; Schiffrin, D. J.; Whyman, R. Synthesis of Thiol-derivatised Gold Nanoparticles in. *J. Chem. Soc. Chem. Commun.* **1994**, *7*, 801–802.

[76] Leff, D. V.; Brandt, L.; Heath, J. R. Synthesis and Characterization of Hydrophobic, Organically-Soluble Gold Nanocrystals Functionalized with Primary Amines. *Langmuir* **1996**, *12*, 4723–4730.

[77] Warner, M. G.; Reed, S. M.; Hutchison, J. E. Small, Water-Soluble, Ligand-Stabilized Gold Nanoparticles Synthesized by Interfacial Ligand Exchange Reactions. *Chem. Mater.* **2000**, *12*, 3316–3320.

[78] Love, J. C.; Estroff, L. A.; Kriebel, J. K.; Nuzzo, R. G.; Whitesides, G. M. Self-Assembled Monolayers of Thiolates on Metals as a Form of Nanotechnology. *Chem. Rev.* **2005**, *105*, 1103-1169.

[79] Woehrle, G. H.; Brown, L. O.; Hutchison, J. E. Thiol-Functionalized, 1.5-nm Gold Nanoparticles through Ligand Exchange Reactions: Scope and Mechanism of Ligand exchange. *J. Am. Chem. Soc.* **2005**, *127*, 2172–2183.

[80] Ojea-Jimenez, I.; Garcia-Fernandez, L.; Lorenzo, J.; Puntes, V. F. Facile Preparation of Cationic Gold Nanoparticle-Bioconjugates for Cell Penetration and Nuclear Targeting. *ACS Nano* **2012**, *6*, 7692–7702.

[81] Bourone, S. D. M.; Kaulen, C.; Homberger, M.; Simon, U. Directed Self-Assembly and Infrared Reflection Absorption Spectroscopy Analysis of Amphiphilic and Zwitterionic Janus Gold Nanoparticles. *Langmuir* **2016**, *32*, 954–962.

[82] Hassinen, J.; Liljeström, V.; Kostiainen, M. A.; Ras, R. H. A. Rapid Cationization of Gold Nanoparticles by Two-Step Phase Transfer. *Angew. Chem. Int. Ed.* **2015**, *54*, 7990–7993.

[83] Pazos-Perez, N.; Abajo, F. J. G. De; Fery, A.; Alvarez-Puebla, R. A. From Nano to Micro: Synthesis and Optical Properties of Homogeneous Spheroidal Gold Particles and Their Superlattices. *Langmuir* **2012**, 8909–8914.

[84] Edwards, P. P.; Thomas, J. M. Gold in a Metallic Divided State - From Faraday to Present-Day Nanoscience. *Angew. Chem. Int. Ed.* **2007**, *46*, 5480–5486.

[85] Lu, X.; Rycenga, M.; Skrabalak, S. E.; Wiley, B.; Xia, Y. Chemical Synthesis of Novel Plasmonic Nanoparticles. *Annu. Rev. Phys. Chem.* **2009**, *60*, 167–192.

[86] Amendola, V.; Bakr, O. M.; Stellacci, F. A study of the Surface Plasmon Resonance of Silver Nanoparticles by the Discrete Dipole Approximation Method: Effect of Shape, Size, Structure, and Assembly. *Plasmonics* **2010**, *5*, 85–97.

[87] Guo, W.; Pleixats, R.; Shafir, A. Water-Soluble Gold Nanoparticles : From Catalytic Selective Nitroarene Reduction in Water to Refractive Index Sensing. *Chem. - Asian J.* **2015**, *10*, 2437-2443.

[88] Chen, H.; Kou, X.; Yang, Z.; Ni, W.; Wang, J. Shape- and Size-Dependent Refractive Index Sensitivity of Gold Nanoparticles. *Langmui* **2008**, *24*, 5233–5237.

[89] Matricardi, C.; Hanske, C.; Garcia-Pomar, J. L.; Langer, J.; Mihi, A.; Liz-Marzán, L. M. Gold Nanoparticle Plasmonic Superlattices as Surface Enhanced Raman Spectroscopy Substrates. *ACS Nano* **2018**, *12*, 8531–8539.

[90] Alaeian, H.; Dionne, J. A. Plasmon nanoparticle superlattices as optical-frequency magnetic metamaterials. *Opt. Express* **2012**, *20*, 15781–15796.

[91] Lal, S.; Link, S.; Halas, N. J. Nano-optics from sensing to waveguiding. *Nat. Photonics* **2007**, *1*, 641–648.

[92] Kelly, K. L.; Coronado, E.; Zhao, L. L.; Schatz, G. C. The Optical Properties of Metal Nanoparticles: The Influence of Size, Shape, and Dielectric Environment. *J. Phys. Chem. B* **2003**, *107*, 668–677.

[93] Lee, J.-H.; Cho, H.-Y.; Choi, H.; Lee, J.-Y.; Choi, J.-W. Application of Gold Nanoparticle to Plasmonic Biosensors. *Int. J. Mol. Sci.* **2018**, *19*, 2021.

[94] Lee, J. H.; Kim, B. C.; Oh, B. K.; Choi, J. W. Highly sensitive localized surface plasmon resonance immunosensor for label-free detection of HIV-1. *Nanomed. - Nanotechnol.* **2013**, *9*, 1018–1026.

[95] Anker, J. N.; Hall, W. P.; Lyandres, O.; Shah, N. C.; Zhao, J.; van Duyne, R. P. Biosensing with plasmonic nanosensors. *Nat. Mater.* **2008**, *7*, 442–453.

[96] Sönnichsen, C.; Reinhard, B. M.; Liphardt, J.; Alivisatos, A. P. A molecular ruler based on plasmon coupling of single gold and silver nanoparticles. *Nat. Biotechnol.* **2005**, *23*, 741–745.

[97] Wu, L.; Wang, Z.; Shen, B. Large-scale gold nanoparticle superlattice and its SERS properties for the quantitative detection of toxic carbaryl. *Nanoscale* **2013**, *5*, 5274–5278.

[98] Maier, S. A.; Kik, P. G.; Atwater, H. A.; Meltzer, S.; Harel, E.; Koel, B. E.; Requicha, A. A. G. Local detection of electromagnetic energy transport below the diffraction limit in metal nanoparticle plasmon waveguides. *Nat. Mater.* **2003**, *2*, 229–232.

[99] Barrow, S. J.; Wei, X.; Baldauf, J. S.; Funston, A. M.; Mulvaney, P. The surface plasmon modes of self-assembled gold nanocrystals. *Nat. Commun.* **2012**, *3*, 1275.

[100] Polman, A.; Atwater, H. A. Plasmonics: Optics at the nanoscale. *Mater. Today* **2005**, *8*, 56.

[101] Zia, R.; Schuller, J. A.; Chandran, A.; Brongersma, M. L. Plasmonics: the next chip-scale technology. *Mater. Today* **2006**, *9*, 20–27.

[102] Miller, D. A. B. Rationale and Challenges for Optical Interconnects to Electronic Chips. *Proc. IEEE* **2000**, *88*, 728–749.

[103] Chen, G.; Chen, H.; Haurylau, M.; Nelson, N.; Albonesi, D.; Fauchet, P. M.; Friedman, E. G. Predictions of CMOS Compatible On-Chip Optical Interconnect. *Integration* **2007**, *40*, 434–446.

[104] Keiser, G. Optical Fiber Communications. in *Wiley Encyclopedia of Telecommunications* (John Wiley & Sons, Inc., 2003).

[105] Ito, T.; Okazaki, S. Pushing the limits of lithography. *Nature* **2000**, *406*, 1027–1031.

[106] Grzelczak, M.; Vermant, J.; Furst, E. M.; Liz-Marzán, L. M. Directed Self-Assembly of Nanoparticles. *ACS Nano* **2010**, *4*, 3591–3605.

[107] Talapin, B. D. V; Shevchenko, E. V; Kornowski, A.; Gaponik, N.; Haase, M.; Rogach, A. L.; Weller, H. A New Approach to Crystallization of CdSe Nanoparticles into Ordered Three-Dimensional. *Adv. Mater.* **2001**, *13*, 1868–1871.

[108] Murray, C. B.; Kagan, C. R.; Bawendi, M. G. Self-Organization of CdSe Nanocrystallites into Three-Dimensional Quantum Dot Superlattices. *Science* **1995**, *270*, 1335–13380.

[109] Kiely, C. J.; Fink, J.; Brust, M.; Bethell, D.; Schiffrin, D. J. Spontaneous ordering of bimodal ensembles of nanoscopic gold clusters. *Nature* **1998**, *396*, 444–446.

[110] Pauling, L. The Principles Determing the Structure of Complex Ionic Crystals. *J. Am. Chem. Soc.* **1929**, *51*, 1010–1026.

[111] Murray, M. J.; Sanders, J. V. Close-packed structures of spheres of two different sizes II . The packing densities of likely arrangements. *Philos. Mag. A* **1980**, *42*, 721–740.

[112] Sanders, J. V; Murray, M. J. Ordered arrangements of spheres of two different sizes in opal. *Nature* **1978**, *275*, 201–203.

[113] Kiely, C. J.; Fink, J.; Zheng, J. G.; Brust, M.; Bethell, D.; Schiffrin, D. J. Ordered Colloidal Nanoalloys. *Adv. Mater.* **2000**, *12*, 640–643.

[114] Redl, F. X.; Cho, K.-S.; Murray, C. B.; O'Brien, S. Three-dimensional binary superlattices of magnetic nanocrystals and semiconductor quantum dots. *Nature* **2003**, *423*, 968–971.

[115] Saunders, A. E.; Korgel, B. A. Observation of an AB Phase in Bidisperse Nanocrystal Superlattices. *ChemPhysChem* **2005**, *6*, 61–65.

[116] Leunissen, M.; Christova, C.; Hynninen, A.; Royall, C.; Campbell, A.; Imhof, A.; Dijkstra, M.; van Roij, R.; van Blaaderen, A. Ionic colloidal crystals of oppositely charged particles. *Nature* **2005**, *437*, 235–240.

[117] Kalsin, A. M.; Fialkowski, M.; Paszewski, M.; Smoukov, S. K.; Bishop, K. J. M.; Grzybowski, B. A. Electrostatic Self-Assembly of Binary Nanoparticle Crystals with a Diamond-Like Lattice. *Science* **2006**, *312*, 420–424.

[118] Ma, N.; Minevich, B.; Liu, J.; Ji, M.; Tian, Y.; Gang, O. Directional Assembly of Nanoparticles by DNA Shapes: Towards Designed Architectures and Functionality. *Top. Curr. Chem.* **2020**, *378*, 1–34.

[119] Gabrys, P. A.; Zornberg, L. Z.; Macfarlane, R. J. Programmable Atom Equivalents: Atomic Crystallization as a Framework for Synthesizing Nanoparticle Superlattices. *Small* **2019**, *1805424*, 1–20.

[120] Alivisatos, A. P.; Johnssont, K. P.; Peng, X.; Wilsont, T. E.; Lowetht, C. J.; Jr, M. P. B.; Schultz, P. G. Organization of 'nanocrystal molecules' using DNA. *Nature* **1996**, *382*, 609–611.

[121] Macfarlane, R. J.; Lee, B.; Jones, M. R.; Harris, N.; Schatz, G. C.; Mirkin, C. A. Nanoparticle Superlattice Engineering with DNA. *Science* **2011**, *334*, 204–208.

[122] Lin, C.; Liu, Y.; Rinker, S.; Yan, H. DNA Tile Based Self-Assembly: Building Complex Nanoarchitectures. *ChemPhysChem* **2006**, *7*, 1641–1647.

[123] Ma, R.-I.; Kallenbach, N. R.; Sheardy, R. D.; Petrillo, M. L.; Seeman, N. C. Three-arm nucleic acid junctions are flexible. *Nucleic Acids Res.* **1986**, *14*, 9745–9753.

[124] Chen, J.; Seeman, N. C. Synthesis from DNA of a molecule with the connectivity of a cube. *Nature* **1991**, *350*, 631–633.

[125] Zhang, Y.; Seeman, N. C. Construction of a DNA-Truncated Octahedron. *J. Am. Chem. Soc.* **1994**, *116*, 1661–1669.

[126] Wang, Y.; Mueller, J. E.; Kemper, B.; Seeman, N. C. Assembly and Characterization of Five-Arm and Six-Arm DNA Branched Junctions. *Biochemistry* **1991**, *30*, 5667–5674.

[127] Wang, X.; Seeman, N. C. Assembly and Characterization of 8-arm and 12-arm DNA Branched Junctions. *J. Am. Chem. Soc.* **2007**, *129*, 8169–8176.

[128] Fu, T.-J.; Seeman, N. C. DNA Double-Crossover Molecules. *Biochemistry* **1993**, *32*, 3211–3220.

[129] Winfree, E.; Liu, F.; Wenzler, L. A.; Seeman, N. C. Design and self-assembly of two-dimensional DNA crystals. *Nature* **1998**, *394*, 539–544.

[130] He, Y.; Tian, Y.; Ribbe, A. E.; Mao, C. Highly Connected Two-Dimensional Crystals of DNA Six-Point-Stars. *J. Am. Chem. Soc.* **2006**, *128*, 15978–15979.

[131] He, Y.; Chen, Y.; Liu, H.; Ribbe, A. E.; Mao, C. Self-Assembly of Hexagonal DNA Two-Dimensional (2D) Arrays. *J. Am. Chem. Soc.* **2005**, *127*, 12202–12203.

[132] Yan, H.; Park, S. H.; Finkelstein, G.; Reif, J. H.; LaBean, T. H. DNA-Templated Self-Assembly of Protein Arrays and Highly Conductive Nanowires. *Science* **2003**, *301*, 1882–1884.

[133] Ding, B.; Sha, R.; Seeman, N. C. Pseudohexagonal 2D DNA Crystals from Double Crossover Cohesion. *J. Am. Chem. Soc.* **2004**, *126*, 10230–10231.

[134] Zheng, J.; Constantinou, P. E.; Micheel, C.; Alivisatos, A. P.; Kiehl, R. A.; Seeman, N. C. Two-Dimensional Nanoparticle Arrays Show the Organizational Power of Robust DNA Motifs. *Nano Lett.* **2006**, *6*, 1502–1504.

[135] Saccà, B.; Niemeyer, C. M. DNA Origami: The Art of Folding DNA. *Angew. Chem. Int. Ed.* **2012**, *51*, 58–66.

[136] Rothemund, P. W. K. Folding DNA to create nanoscale shapes and patterns. *Nature* **2006**, *440*, 297–302.

[137] Andersen, E. S.; Dong, M.; Nielsen, M. M.; Jahn, K.; Lind-Thomsen, A.; Mamdouh, W.; Gothelf, K. V.; Besenbacher, F.; Kjems, J. DNA Origami Design of Dolphin-Shaped Structures with Flexible Tails. *ACS Nano* **2008**, *2*, 1213–1218.

[138] Douglas, S. M.; Marblestone, A. H.; Teerapittayanon, S.; Vazquez, A.; Church, G. M.; Shih, W. M. Rapid prototyping of 3D DNA-origami shapes with caDNAno. *Nucleic Acids Res.* **2009**, *37*, 5001–5006.

[139] Tian, Y.; Wang, T.; Liu, W.; Xin, H. L.; Li, H.; Ke, Y.; Shih, W. M.; Gang, O. Prescribed nanoparticle cluster architectures and low-dimensional arrays built using octahedral DNA origami frames. *Nat. Nanotechnol.* **2015**, *10*, 637–644.

[140] Liu, W.; Tagawa, M.; Xin, H. L.; Wang, T.; Emamy, H.; Li, H.; Yager, K. G.; Starr, F. W.; Tkachenko, A. V; Gang, O. Diamond family of nanoparticle superlattices. *Science* **2016**, *351*, 582–586.

[141] Wang, M.; Dai, L.; Duan, J.; Ding, Z.; Wang, P.; Li, Z.; Xing, H.; Tian, Y. Programmable Assembly of Nano-architectures through Designing Anisotropic DNA Origami Patches. *Angew. Chem. Int. Ed.* **2020**, *59*, 6389–6396.

[142] Jin, R.; Zeng, C.; Zhou, M.; Chen, Y. Atomically Precise Colloidal Metal Nanoclusters and Nanoparticles: Fundamentals and Opportunities. *Chem. Rev.* **2016**, *116*, 10346–10413.

[143] Service, R. F. The Next Big(ger) Thing. *Science* **2012**, *335*, 1167–1167.

[144] Uchida, M.; Klem, M. T.; Allen, M.; Suci, P.; Flenniken, M.; Gillitzer, E.; Varpness, Z.; Liepold, L. O.; Young, M.; Douglas, T. Biological Containers: Protein Cages as Multifunctional Nanoplatforms. *Adv. Mater.* **2007**, *19*, 1025–1042.

[145] Aumiller, W. M.; Uchida, M.; Douglas, T. Protein cage assembly across multiple length scales. *Chem. Soc. Rev.* **2018**, *47*, 3433–3469.

[146] Stasiak, A. Z.; Larquet, E.; Stasiak, A.; Müller, S.; Engel, A.; Van Dyck, E.; West, S. C.; Egelman, E. H. The human Rad52 protein exists as a heptameric ring. *Curr. Biol.* **2000**, *10*, 337–340.

[147] Heddle, J. Protein cages, rings and tubes: useful components of future nanodevices? *Nanotechnol. Sci. Appl.* **2008**, *1*, 67–78.

[148] Beck, T.; Tetter, S.; Künzle, M.; Hilvert, D. Construction of Matryoshka-Type Structures from Supercharged Protein Nanocages. *Angew. Chem. Int. Ed.* **2015**, *54*, 937–940.

[149] Khoshnejad, M.; Greineder, C. F.; Pulsipher, K. W.; Villa, C. H.; Altun, B.; Pan, D. C.; Tsourkas, A.; Dmochowski, I. J.; Muzykantov, V. R. Ferritin Nanocages with Biologically Orthogonal Conjugation for Vascular Targeting and Imaging. *Bioconjug. Chem.* **2018**, *29*, 1209–1218.

[150] Domínguez-Vera, J. M.; Colacio, E. Nanoparticles of Prussian Blue Ferritin: A New Route for Obtaining Nanomaterials. *Inorg. Chem.* **2003**, *42*, 6983–6985.

[151] Pulsipher, K. W.; Villegas, J. A.; Roose, B. W.; Hicks, T. L.; Yoon, J.; Saven, J. G.; Dmochowski, I. J. Thermophilic Ferritin 24mer Assembly and Nanoparticle Encapsulation Modulated by Interdimer Electrostatic Repulsion. *Biochemistry* **2017**, *56*, 3596–3606.

[152] Ueno, T.; Suzuki, M.; Goto, T.; Matsumoto, T.; Nagayama, K.; Watanabe, Y. Size-Selective Olefin Hydrogenation by a Pd Nanocluster Provided in an Apo-Ferritin Cage. *Angew. Chem. Int. Ed.* **2004**, *43*, 2527–2530.

[153] Abe, S.; Hirata, K.; Ueno, T.; Morino, K.; Shimizu, N.; Yamamoto, M.; Takata, M.; Yashima, E.; Watanabe, Y. Polymerization of Phenylacetylene by Rhodium Complexes within a Discrete Space of apo-Ferritin. *J. Am. Chem. Soc.* **2009**, *131*, 6958–6960.

[154] Nichols, R. J.; Cassidy-Amstutz, C.; Chaijarasphong, T.; Savage, D. F. Encapsulins: molecular biology of the shell. *Crit. Rev. Biochem. Mol. Biol.* **2017**, *52*, 583–594.

[155] Lai, Y. T.; Reading, E.; Hura, G. L.; Tsai, K. L.; Laganowsky, A.; Asturias, F. J.; Tainer, J. A.; Robinson, C. V.; Yeates, T. O. Structure of a designed protein cage that self-assembles into a highly porous cube. *Nat. Chem.* **2014**, *6*, 1065–1071.

[156] Bale, J. B.; Gonen, S.; Liu, Y.; Sheffler, W.; Ellis, D.; Thomas, C.; Cascio, D.; Yeates, T. O.; Gonen, T.; King, N. P.; Baker, D. Accurate design of megadalton-scale two-component icosahedral protein complexes. *Science* **2016**, *353*, 389–394.

[157] King, N. P.; Bale, J. B.; Sheffler, W.; McNamara, D. E.; Gonen, S.; Gonen, T.; Yeates, T. O.; Baker, D. Accurate design of co-assembling multi-component protein nanomaterials. *Nature* **2014**, *510*, 103–108.

[158] Fletcher, J. M.; Harniman, R. L.; Barnes, F. R. H.; Boyle, A. L.; Collins, A.; Mantell, J.; Sharp, T. H.; Antognozzi, M.; Booth, P. J.; Linden, N.; Miles, M. J.; Sessions, R. B.; Verkade, P.; Woolfson, D. N. Self-Assembling Cages from Coiled-Coil Peptide Modules. *Science* **2013**, *340*, 595–599.

[159] Cristie-David, A. S.; Marsh, E. N. G. Metal-dependent assembly of a protein nano-cage. *Protein Sci.* **2019**, *28*, 1620–1629.

[160] Malay, A. D. *et al.* An ultra-stable gold-coordinated protein cage displaying reversible assembly. *Nature* **2019**, *569*, 438–442.

[161] Golub, E.; Subramanian, R. H.; Esselborn, J.; Alberstein, R. G.; Bailey, J. B.; Chiong, J. A.; Yan, X.; Booth, T.; Baker, T. S.; Tezcan, F. A. Constructing protein polyhedra via orthogonal chemical interactions. *Nature* **2020**, *578*, 172–176.

[162] Lawson, D. M.; Artymiuk, P. J.; Yewdall, S. J.; Smith, J. M. A.; Livingstone, J. C.; Treffry, A.; Luzzago, A.; Levi, S.; Arosio, P.; Cesareni, G.; Thomas, C. D.; Shaw, W. V.; Harrison, P. M. Solving the structure of human H ferritin by genetically engineering intermolecular crystal contacts. *Nature* **1991**, *349*, 541–544.

[163] Crichton, R. R.; Declercq, J. P. X-ray structures of ferritins and related proteins. *Biochim. Biophys. Acta* **2010**, *1800*, 706–718.

[164] Harrison, P. M.; Arosio, P. The ferritins: molecular properties, iron storage function and cellular regulation. *Biochim. Biophys. Acta* **1996**, *1275*, 161–203.

[165] Zhao, G.; Bou-Abdallah, F.; Arosio, P.; Levi, S.; Janus-Chandler, C.; Chasteen, N. D. Multiple Pathways for Mineral Core Formation in Mammalian Apoferritin. The Role of Hydrogen Peroxide. *Biochemistry* **2003**, *42*, 3142–3150.

[166] Chasteen, N. D.; Harrison, P. M. Mineralization in Ferritin: An Efficient Means of Iron Storage. *J. Struct. Biol.* **1999**, *126*, 182–194.

[167] Levi, S.; Santambrogio, P.; Corsi, B.; Cozzi, A.; Arosio, P. Evidence that the specifity of iron incorporation into homopolymers of human ferritin L- and H-chains is conferred by the nucleation and ferroxidase centres. *Biochem. J.* **1996**, *314*, 139–144.

[168] Liu, X.; Jin, W.; Theil, E. C. Opening protein pores with chaotropes enhances Fe reduction and chelation of Fe from the ferritin biomineral. *PNAS* **2003**, *100*, 3653–3658.

[169] Linder, M. C.; Kakavandi, H. R.; Miller, P.; Wirth, P. L.; Nagel, G. M. Dissociation of Ferritins. *Arch. Biochem. Biophys.* **1989**, *269*, 485–496.

[170] Maity, B.; Fujita, K.; Ueno, T. Use of the confined spaces of apo-ferritin and virus capsids as nanoreactors for catalytic reactions. *Curr. Opin. Chem. Biol.* **2015**, *25*, 88–97.

[171] Niemeyer, J.; Abe, S.; Hikage, T.; Ueno, T.; Erker, G.; Watanabe, Y. Noncovalent insertion of ferrocenes into the protein shell of apo-ferritin. *Chem. Commun.* **2008**, *1*, 6519–6521.

[172] Maity, B.; Fukumori, K.; Abe, S.; Ueno, T. Immobilization of two organometallic complexes into a single cage to construct protein-based microcompartments. *Chem. Commun.* **2016**, *52*, 5463-5466.

[173] Cutrin, J. C.; Crich, S. G.; Burghelea, D.; Dastru, W.; Aime, S. Curcumin/Gd Loaded Apoferritin: A Novel 'Theranostic' Agent to Prevent Hepatocellular Damage in Toxic Induced Acute Hepatitis. *Mol. Pharm.* **2013**, *10*, 2079–2085.

[174] Khoshnejad, M.; Parhiz, H.; Shuvaev, V. V.; Dmochowski, I. J.; Muzykantov, V. R. Ferritin-based drug delivery systems: Hybrid nanocarriers for vascular immunotargeting. *J. Control. Release* **2018**, *282*, 13–24.

[175] Swift, J.; Wehbi, W. A.; Kelly, B. D.; Stowell, X. F.; Saven, J. G.; Dmochowski, I. J. Design of Functional Ferritin-Like Proteins with Hydrophobic Cavities. *J. Am. Chem. Soc.* **2006**, *128*, 6611-6619.

[176] Wang, Z.; Takezawa, Y.; Aoyagi, H.; Abe, S.; Hikage, T.; Watanabe, Y.; Kitagawa, S.; Ueno, T. Definite coordination arrangement of organometallic palladium complexes accumulated on the designed interior surface of apo-ferritin. *Chem. Commun.* **2011**, *47*, 170–172.

[177] Abe, S.; Niemeyer, J.; Abe, M.; Takezawa, Y.; Ueno, T.; Hikage, T.; Erker, G.; Watanabe, Y. Control of the Coordination Structure of Organometallic Palladium Complexes in an apo-Ferritin Cage. *J. Am. Chem. Soc.* **2008**, *130*, 10512–10514.

[178] Zhang, S.; Zang, J.; Zhang, X.; Chen, H.; Mikami, B.; Zhao, G. “Silent” Amino Acid Residues at Key Subunit Interfaces Regulate the Geometry of Protein Nanocages. *ACS Nano* **2016**, *10*, 10382–10388.

[179] Chen, H.; Zhang, S.; Xu, C.; Zhao, G. Engineering protein interfaces yields ferritin disassembly and reassembly under benign experimental conditions. *Chem. Commun.* **2016**, *52*, 7402–7405.

[180] Lee, L. A.; Niu, Z.; Wang, Q. Viruses and Virus-Like Protein Assemblies-Chemically Programmable Nanoscale Building Blocks. *Nano Res.* **2009**, *2*, 349–364.

[181] Wong, K. K. W.; Whilton, N. T.; Cölfen, H.; Douglas, T.; Mann, S. Hydrophobic proteins: Synthesis and characterisation of organic-soluble alkylated ferritins. *Chem. Commun.* **1998**, 1621–1622.

[182] McPherson, A.; Gavira, J. A. Introduction to protein crystallization. *Acta Crystallogr.* **2014**, *70*, 2-20.

[183] Berry, I. M.; Dym, O.; Esnouf, R. M.; Harlos, K.; Meged, R.; Perrakis, A.; Sussman, J. L.; Walter, T. S.; Wilson, J.; Messerschmidt, A. SPINE high-throughput crystallization, crystal imaging and recognition techniques: current state, performance analysis, new technologies and future. *Acta Crystallogr. Sect. D* **2006**, *62*, 1137–1149.

[184] Granier, T.; Gallois, B.; Dautant, A.; D'Estaintot, B. L.; Précigoux, G. Preliminary X-ray diffraction studies of the tetragonal form of native horse-spleen apoferritin. *Acta Crystallogr. Sect. D* **1996**, *52*, 594–596.

[185] Granier, T.; Gallois, B.; Dautant, A.; Langlois D'Estaintot, B.; Precigoux, G. Comparison of the Structures of the Cubic and Tetragonal Forms of Horse-Spleen Apoferritin. *Acta Crystallogr. Sect. D* **1997**, *53*, 580–587.

[186] Sontz, P. A.; Bailey, J. B.; Ahn, S.; Tezcan, F. A. A Metal Organic Framework with Spherical Protein Nodes: Rational Chemical Design of 3D Protein Crystals. *J. Am. Chem. Soc.* **2015**, *137*, 11598–11601.

[187] Bailey, J. B.; Zhang, L.; Chiong, J. A.; Ahn, S.; Tezcan, F. A. Synthetic Modularity of Protein-Metal-Organic Frameworks. *J. Am. Chem. Soc.* **2017**, *139*, 8160–8166.

[188] Liljeström, V.; Seitsonen, J.; Kostiainen, M. A. Electrostatic Self-Assembly of Soft Matter Nanoparticle Co-Crystals with Tunable Lattice Parameters. *ACS Nano* **2015**, *9*, 11278–11285.

[189] Lach, M. J. Master thesis: Ordered assembly of two oppositely charged ferritin containers and characterization of a binary structure by X-ray crystallography. (2016).

[190] Kostiainen, M. A.; Hiekkataipale, P.; Laiho, A.; Lemieux, V.; Seitsonen, J.; Ruokolainen, J.; Ceci, P. Electrostatic assembly of binary nanoparticle superlattices using protein cages. *Nat. Nanotechnol.* **2012**, *8*, 52–56.

[191] Maity, B.; Abe, S.; Ueno, T. Observation of gold sub-nanocluster nucleation within a crystalline protein cage. *Nat. Commun.* **2017**, *8*, 14820.

[192] Beck, T.; Krasauskas, A.; Gruene, T.; Sheldrick, G. M. A magic triangle for experimental phasing of macromolecules. *Acta Crystallogr. Sect. D* **2008**, *64*, 1179–1182.

[193] Lawrence, M. S.; Phillips, K. J.; Liu, D. R. Supercharging Proteins Can Impart Unusual Resilience. *J. Am. Chem. Soc.* **2007**, *129*, 10110–10112.

[194] Miklos, A. E. *et al.* Structure-Based Design of Supercharged, Highly Thermoresistant Antibodies. *Chem. Biol.* **2012**, *19*, 449–455.

[195] Leaver-Fay, A. *et al.* Rosetta3: An Object-Oriented Software Suite for the Simulation and Design of Macromolecules. *Methods Enzym.* **2011**, *487*, 545–574.

[196] Liu, Y.; Kuhlman, B. RosettaDesign server for protein design. *Nucleic Acids Res.* **2006**, *34*, 235-238.

[197] Okuda, M.; Eloi, J.-C.; Ward Jones, S. E.; Sarua, A.; Richardson, R. M.; Schwarzacher, W. Fe3O4 nanoparticles: protein-mediated crystalline magnetic superstructures. *Nanotechnology* **2012**, *23*, 415601-41567.

[198] Tsukamoto, R.; Iwahori, K.; Muraoka, M.; Yamashita, I. Synthesis of Co3O4 Nanoparticles Using the Cage-Shaped Protein, Apoferritin. *Bull. Chem. Soc. Jpn.* **2005**, *78*, 2075–2081.

[199] Tsukamoto, R.; Muraoka, M.; Fukushige, Y.; Nakagawa, H.; Kawaguchi, T.; Nakatsuji, Y.; Yamashita, I. Improvement of Co3O4 Nanoparticle Synthesis in Apoferritin Cavity by Outer Surface PEGylation. *Bull. Chem. Soc. Jpn.* **2008**, *81*, 1669–1674.

[200] Okuda, M.; Suzumoto, Y.; Yamashita, I. Bioinspired Synthesis of Homogenous Cerium Oxide Nanoparticles and Two- or Three-Dimensional Nanoparticle Arrays Using Protein Supramolecules. *Cryst. Growth Des.* **2011**, *11*, 2540–2545.

[201] Künzle, M.; Eckert, T.; Beck, T. Metal-Assisted Assembly of Protein Containers Loaded with Inorganic Nanoparticles. *Inorg. Chem.* **2018**, *57*, 13431–13436.

[202] Wu, W.; He, Q.; Jiang, C. Magnetic Iron Oxide Nanoparticles: Synthesis and Surface Functionalization Strategies. *Nanoscale Res. Lett.* **2008**, *3*, 397–415.

[203] Dong, J.; Song, L.; Yin, J. J.; He, W.; Wu, Y.; Gu, N.; Zhang, Y. Co3O4 Nanoparticles with Multi-Enzyme Activities and Their Application in Immunohistochemical Assay. *ACS Appl. Mater. Inter* **2014**, *6*, 1959–1970.

[204] Xu, C.; Qu, X. Cerium oxide nanoparticle: a remarkably versatile rare earth nanomaterial for biological applications. *NPG Asia Mater.* **2014**, *6*, e90.

[205] Hennequin, B.; Turyanska, L.; Ben, T.; Beltrán, A. M.; Molina, S. I.; Li, M.; Mann, S.; Patanè, A.; Thomas, N. R. Aqueous Near-Infrared Fluorescent Composites Based on Apoferritin-Encapsulated PbS Quantum Dots. *Adv. Mater.* **2008**, *20*, 3592–3596.

[206] Fukano, H.; Takahashi, T.; Aizawa, M.; Yoshimura, H. Synthesis of Uniform and Dispersive Calcium Carbonate Nanoparticles in a Protein Cage through Control of Electrostatic Potential. *Inorg. Chem.* **2011**, *50*, 6526–6532.

[207] Fan, R.; Chew, S. W.; Cheong, V. V.; Orner, B. P. Fabrication of Gold Nanoparticles inside Unmodified Horse Spleen Apoferritin. *Small* **2010**, *6*, 1483–1487.

[208] Kim, M.; Rho, Y.; Jin, K. S.; Ahn, B.; Jung, S.; Kim, H.; Ree, M. pH-Dependent Structures of Ferritin and Apoferritin in Solution: Disassembly and Reassembly. *Biomacromolecules* **2011**, *12*, 1629–1640.

[209] Künzle, M.; Mangler, J.; Lach, M.; Beck, T. Peptide - directed encapsulation of inorganic nanoparticles into protein containers. *Nanoscale* **2018**, *10*, 22917–22926.

[210] Liu, A.; Verwegen, M.; De Ruiter, M. V.; Maassen, S. J.; Traulsen, C. H. H.; Cornelissen, J. J. L. M. Protein Cages as Containers for Gold Nanoparticles. *J. Phys. Chem. B* **2016**, *120*, 6352–6357.

[211] Capehart, S. L.; Coyle, M. P.; Glasgow, J. E.; Francis, M. B. Controlled Integration of Gold Nanoparticles and Organic Fluorophores using Synthetically Modified MS2 Viral Capsids. *J. Am. Chem. Soc.* **2013**, *135*, 3011–3016.

[212] Spinelli, P.; Hebbink, M.; De Waele, R.; Black, L.; Lenzmann, F.; Polman, A. Optical Impedance Matching Using Coupled Plasmonic Nanoparticle Arrays. *Nano Lett.* **2011**, *11*, 1760–1765.

[213] Stratakis, E.; Kymakis, E. Nanoparticle-based plasmonic organic photovoltaic devices. *Mater. Today* **2013**, *16*, 133–146.

[214] Rahme, K.; Nolan, M. T.; Doody, T.; McGlacken, G. P.; Morris, M. A.; O'Driscoll, C.; Holmes, J. D. Highly stable PEGylated gold nanoparticles in water: applications in biology and catalysis. *RSC Adv.* **2013**, *3*, 21016–21024.

[215] Logunov, S. L.; Ahmadi, T. S.; El-Sayed, M. A.; Khoury, J. T.; Whetten, R. L. Electron Dynamics of Passivated Gold Nanocrystals Probed by Subpicosecond Transient Absorption Spectroscopy. *J. Phys. Chem. B* **1997**, *101*, 3713–3719.

[216] Alvarez, M. M.; Khoury, J. T.; Schaaff, T. G.; Shafigullin, M. N.; Vezmar, I.; Whetten, R. L. Optical Absorption Spectra of Nanocrystal Gold Molecules. *J. Phys. Chem. B* **1997**, *101*, 3706–3712.

[217] Sivaraman, S. K.; Kumar, S.; Santhanam, V. Monodisperse sub-10nm gold nanoparticles by reversing the order of addition in Turkevich method - The role of chloroauric acid. *J. Colloid Interf. Sci.* **2011**, *361*, 543–547.

[218] Frens, G. Controlled Nucleation for the Regulation of the Particle Size in Monodisperse Gold Suspensions. *Nat. Phys. Sci.* **1973**, *241*, 20–22.

[219] Kimling, J.; Maier, M.; Okenve, B.; Kotaidis, V.; Ballot, H.; Plech, A. Turkevich Method for Gold Nanoparticle Synthesis Revisited. *J. Phys. Chem. B* **2006**, *110*, 15700–15707.

[220] Kumar, S.; Gandhi, K. S.; Kumar, R. Modeling of Formation of Gold Nanoparticles by Citrate Method. *Ind. Eng. Chem. Res.* **2007**, *46*, 3128–3136.

[221] Brust, M.; Fink, J.; Bethell, D.; Schiffrin, D. J.; Kiely, C. Synthesis and Reactions of Functionalised Gold Nanoparticles. *J. Chem. Soc. Chem. Commun.* **1995**, 1655–1656.

[222] Goulet, P. J. G.; Lennox, R. B. New Insights into Brust−Schiffrin Metal Nanoparticle Synthesis. *J. Am. Chem. Soc.* **2010**, *132*, 9582–9584.

[223] Peng, S.; Lee, Y.; Wang, C.; Yin, H.; Dai, S.; Sun, S. A Facile Synthesis of Monodisperse Au Nanoparticles and Their Catalysis of CO Oxidation. *Nano Res.* **2008**, *1*, 229–234.

[224] Madras, G.; McCoy, B. J. Temperature effects during Ostwald ripening. *J. Chem. Phys.* **2003**, *119*, 1683–1693.

[225] Ackerson, C. J.; Jadzinsky, P. D.; Kornberg, R. D. Thiolate Ligands for Synthesis of Water-Soluble Gold Clusters. *J. Am. Chem. Soc.* **2005**, *127*, 6550–6551.

[226] Pillai, P. P.; Kowalczyk, B.; Kandere-Grzybowska, K.; Borkowska, M.; Grzybowski, B. A. Engineering Gram Selectivity of Mixed-Charge Gold Nanoparticles by Tuning the Balance of Surface Charges. *Angew. Chem. Int. Ed.* **2016**, *55*, 8610–8614.

[227] Frisch, M. *et al.* Gaussian 09 (Revision A02). *Gaussian Inc. Wallingford CT* **2009**,.

[228] Pillai, P. P.; Huda, S.; Kowalczyk, B.; Grzybowski, B. A. Controlled pH Stability and Adjustable Cellular Uptake of Mixed-Charge Nanoparticles. *J. Am. Chem. Soc.* **2013**, *135*, 6392–6395.

[229] Levin, C. S.; Bishnoi, S. W.; Grady, N. K.; Halas, N. J. Determining the Conformation of Thiolated Poly(ethylene glycol) on Au Nanoshells by Surface-Enhanced Raman Scattering Spectroscopic Assay. *Anal. Chem.* **2006**, *78*, 3277–3281.

[230] Peter, S. K.; Kaulen, C.; Hoffmann, A.; Ogieglo, W.; Karthäuser, S.; Homberger, M.; Herres-Pawlis, S.; Simon, U. Stepwise Growth of Ruthenium Terpyridine Complexes on Au Surfaces. *J. Phys. Chem. C* **2019**, *123*, 6537–6548.

[231] Marbella, L. E.; Millstone, J. E. NMR Techniques for Noble Metal Nanoparticles. *Chem. Mater.* **2015**, *27*, 2721–2739.

[232] Wuelfing, W. P.; Gross, S. M.; Miles, D. T.; Murray, R. W. Nanometer Gold Clusters Protected by Surface-Bound Monolayers of Thiolated Poly(ethylene glycol) Polymer Electrolyte. *J. Am. Chem. Soc.* **1998**, *120*, 12696–12697.

[233] Raman, C. V.; Krishnan, K. S. The Negative Absorption of Radiation. *Nature* **1928**, *122*, 12–13.

[234] Nie, S.; Emory, S. R. Probing Single Molecules and Single Nanoparticles by Surface-Enhanced Raman Scattering. *Science* **1997**, *275*, 1102–1106.

[235] Blackie, E. J.; Le Ru, E. C.; Etchegoin, P. G. Single-Molecule Surface-Enhanced Raman Spectroscopy of Nonresonant Molecules. *J. Am. Chem. Soc.* **2009**, *131*, 14466–14472.

[236] I.I. Rabi. Space Quantization in a Gyrating Magnetic Field. *Phys. Rev.* **1937**, *51*, 652–654.

[237] Rabi, I. I.; Zacharias, J. R.; Millman, S.; Kusch, P. A New Method of Measuring Nuclear Magnetic Moment. *Phys. Rev.* **1938**, *53*, 318.

[238] Hostetler, M. J.; Wingate, J. E.; Zhong, C.-J.; Harris, J. E.; Vachet, R. W.; Clark, M. R.; Londono, J. D.; Green, S. J.; Stokes, J. J.; Wignall, G. D.; Glish, G. L.; Porter, M. D.; Evans, N. D.; Murray, R. W. Alkanethiolate Gold Cluster Molecules with Core Diameters from 1.5 to 5.2 nm: Core and Monolayer Properties as a Function of Core Size. *Langmuir* **1998**, *14*, 17–30.

[239] Song, Y.; Harper, A. S.; Murray, R. W. Ligand Heterogeneity on Monolayer-Protected Gold Clusters. *Langmuir* **2005**, *21*, 5492–5500.

[240] Badia, A.; Gao, W.; Singh, S.; Demers, L.; Cuccia, L.; Reven, L. Structure and Chain Dynamics of Alkanethiol-Capped Gold Colloids. *Langmuir* **1996**, *12*, 1262–1269.

[241] Liu, X.; Yu, M.; Kim, H.; Mameli, M.; Stellacci, F. Determination of monolayer-protected gold nanoparticle ligand-shell morphology using NMR. *Nat. Commun.* **2012**, *3*, 1182–1189.

[242] Smith, A. M.; Marbella, L. E.; Johnston, K. A.; Hartmann, M. J.; Crawford, S. E.; Kozycz, L. M.; Seferos, D. S.; Millstone, J. E. Quantitative Analysis of Thiolated Ligand Exchange on Gold Nanoparticles Monitored by 1H NMR Spectroscopy. *Anal. Chem.* **2015**, *87*, 2771–2778.

[243] Ogg, R. J.; Kingsley, P. B.; Taylor, J. S. WET, a T1- and B1-Insensitive Water-Suppression Method for in Vivo Localized 1H NMR Spectroscopy. *Journal of Magnetic Resonance, Series B* **1994**, *104*, 1–10.

[244] Kaulen, C.; Homberger, M.; Bourone, S.; Babajani, N.; Karthäuser, S.; Besmehn, A.; Simon, U. Differential Adsorption of Gold Nanoparticles to Gold/Palladium and Platinum Surfaces. *Langmuir* **2014**, *30*, 574–583.

[245] Kastilani, R.; Wong, R.; Pozzo, L. D. Efficient Electrosteric Assembly of Nanoparticle Heterodimers and Linear Heteroassemblies. *Langmuir* **2018**, *34*, 826–836.

[246] Smith, A. M.; Johnston, K. A.; Crawford, S. E.; Marbella, L. E.; Millstone, J. E. Ligand density quantification on colloidal inorganic nanoparticles. *Analyst* **2017**, *142*, 11–29.

[247] Bhattacharjee, S. DLS and zeta potential - What they are and what they are not? *J. Control. Release* **2016**, *235*, 337–351.

[248] Missana, T.; Adell, A. On the Applicability of DLVO Theory to the Prediction of Clay Colloids Stability. *J. Colloid Interf. Sci.* **2000**, *230*, 150–156.

[249] Agnihotri, S.; Mukherji, S.; Mukherji, S. Size-controlled silver nanoparticles synthesized over the range 5–100 nm using the same protocol and their antibacterial efficacy. *RSC Adv.* **2014**, *4*, 3974–3983.

[250] Helmlinger, J.; Sengstock, C.; Groß-Heitfeld, C.; Mayer, C.; Schildhauer, T. A.; Köller, M.; Epple, M. Silver nanoparticles with different size and shape: Equal cytotoxicity, but different antibacterial effects. *RSC Adv.* **2016**, *6*, 18490–18501.

[251] Heath, J. R.; Knobler, C. M.; Leff, D. V. Pressure/Temperature Phase Diagrams and Superlattices of Organically Functionalized Metal Nanocrystal Monolayers: The Influence of Particle Size, Size Distribution, and Surface Passivant. *J. Phys. Chem. B* **1997**, *5647*, 189–197.

[252] Yamamoto, M.; Nakamoto, M. Novel preparation of monodispersed silver nanoparticles via amine adducts derived from insoluble silver myristate in tertiary alkylamine. *J. Mater. Chem.* **2003**, *2*, 10–11.

[253] Yamamoto, M.; Kashiwagi, Y.; Nakamoto, M. Size-Controlled Synthesis of Monodispersed Silver Nanoparticles Capped by Long-Chain Alkyl Carboxylates from Silver Carboxylate and Tertiary Amine. *Langmuir* **2006**, *22*, 8581–8586.

[254] Jeevanandam, P.; Srikanth, C. K.; Dixit, S. Synthesis of monodisperse silver nanoparticles and their self-assembly through simple thermal decomposition approach. *Mater. Chem. Phys.* **2010**, *122*, 402–407.

[255] Park, J.; Kwon, S. G.; Jun, S. W.; Kim, B. H.; Hyeon, T. Large-Scale Synthesis of Ultra-Small-Sized Silver Nanoparticles. *ChemPhysChem* **2012**, *13*, 2540–2543.

[256] Hao, C.; Wang, D.; Zheng, W.; Peng, Q. Growth and assembly of monodisperse Ag nanoparticles by exchanging the organic capping ligands. *J. Mater. Res.* **2009**, *24*, 352–356.

[257] Abe, S.; Hikage, T.; Watanabe, Y.; Kitagawa, S.; Ueno, T. Mechanism of Accumulation and Incorporation of Organometallic Pd Complexes into the Protein Nanocage of apo-Ferritin. *Inorg. Chem.* **2010**, *49*, 6967–6973.

[258] Pontillo, N.; Pane, F.; Messori, L.; Amoresano, A.; Merlino, A. Cisplatin encapsulation within a ferritin nanocage: a high-resolution crystallographic study. *Chem. Commun.* **2016**, *52*, 4136-4139.

[259] Liu, G.; Wang, J.; Lea, S. A.; Lin, Y. Bioassay Labels Based on Apoferritin Nanovehicles. *ChemBioChem* **2006**, *7*, 1315–1319.

[260] Huang, P.; Rong, P.; Jin, A.; Yan, X.; Zhang, M. G.; Lin, J.; Hu, H.; Wang, Z.; Yue, X.; Li, W.; Niu, G.; Zeng, W.; Wang, W.; Zhou, K.; Chen, X. Dye-Loaded Ferritin Nanocages for Multimodal Imaging and Photothermal Therapy. *Adv. Mater.* **2014**, *26*, 6401–6408.

[261] Webb, B.; Frame, J.; Zhao, Z.; Lee, M. L.; Watt, G. D. Molecular Entrapment of Small Molecules within the Interior of Horse Spleen Ferritin. *Arch. Biochem. Biophys.* **1994**, *309*, 178–183.

[262] Rodarte, A. L.; Tao, A. R. Plasmon − Exciton Coupling between Metallic Nanoparticles and Dye Monomers. *J. Phys. Chem. C* **2017**, *121*, 3496–3502.

[263] Luo, Y.; Zhao, J. Plasmon-exciton interaction in colloidally fabricated metal nanoparticle-quantum emitter nanostructures. *Nano Res.* **2019**, *12*, 2164–2171.

[264] Okuda, M.; Schwarze, T.; Eloi, J.-C.; Ward Jones, S. E.; Heard, P. J.; Sarua, A.; Ahmad, E.; Kruglyak, V. V; Grundler, D.; Schwarzacher, W. Top-down design of magnonic crystals from bottom-up magnetic nanoparticle through protein array. *Nanotechnology* **2017**, *28*, 155301.

[265] Sugiyama, S.; Maruyama, M.; Sazaki, G.; Hirose, M.; Adachi, H.; Takano, K.; Murakami, S.; Inoue, T.; Mori, Y.; Matsumura, H. Growth of Protein Crystals in Hydrogels Prevents Osmotic Shock. *J. Am. Chem. Soc.* **2012**, *134*, 5786–5789.

[266] Chayen, N. E.; Shaw Stewart, P. D.; Blow, D. M. Microbatch crystallization under oil - a new technique allowing many small-volume crystallization trials. *J. Cryst. Growth* **1992**, *122*, 176-180.

[267] Chayen, N. E. Comparative Studies of Protein Crystallization by Vapour-Diffusion and Microbatch Techniques. *Acta Crystallogr. Sect. D Biol. Crystallogr.* **1998**, *54*, 8–15.

[268] Rayment, I. Small-Scale Batch Crystallization of Proteins Revisited: An Underutilized Way to Grow Large Protein Crystals. *Structure* **2002**, *10*, 147–151.

[269] Yan, E.-K.; Cao, H.-L.; Zhang, C.-Y.; Lu, Q.-Q.; Ye, Y.-J.; He, J.; Huang, L.-J.; Yin, D.-C. Cross-linked protein crystals by glutaraldehyde and their applications. *RSC Adv.* **2015**, *5*, 26163-26174.

[270] Margolin, A. L.; Navia, M. A. Protein Crystals as Novel Catalytic Materials. *Angew. Chem. Int. Ed.* **2001**, *40*, 2204–2222.

[271] Lusty, C. J. A gentle vapor-diffusion technique for cross-linking of protein crystals for cryocrystallography. *J. Appl. Crystallogr.* **1999**, *32*, 106–112.

[272] Bingham, J. M.; Anker, J. N.; Kreno, L. E.; Duyne, R. P. Van. Gas Sensing with High-Resolution Localized Surface Plasmon Resonance Spectroscopy. *J. Am. Chem. Soc.* **2010**, *132*, 17358-17359.

[273] Kumar, A.; Kim, S.; Nam, J.-M. Plasmonically Engineered Nanoprobes for Biomedical Applications. *J. Am. Chem. Soc.* **2016**, *138*, 14509–14525.

[274] Lednicky, T.; Bonyar, A. Large Scale Fabrication of Ordered Gold Nanoparticle–Epoxy Surface Nanocomposites and Their Application as Label-free Plasmonic DNA Biosensors. *ACS Appl. Mater. Inter* **2020**, *12*, 4804–4814.

[275] Li, J.; Deng, T. S.; Liu, X.; Dolan, J. A.; Scherer, N. F.; Nealey, P. F. Hierarchical Assembly of Plasmonic Nanoparticle Heterodimer Arrays with Tunable Sub-5 nm Nanogaps. *Nano Lett.* **2019**, *19*, 4314–4320.

[276] Yun, C. S.; Javier, A.; Jennings, T.; Fisher, M.; Hira, S.; Peterson, S.; Hopkins, B.; Reich, N. O.; Strouse, G. F. Nanometal Surface Energy Transfer in Optical Rulers, Breaking the FRET Barrier. *J. Am. Chem. Soc.* **2005**, *127*, 3115–3119.

[277] Sen, T.; Sadhu, S.; Patra, A. Surface energy transfer from rhodamine 6G to gold nanoparticles: A spectroscopic ruler. *Appl. Phys. Lett.* **2007**, *91*, 1–4.

[278] Park, S. Y.; Lee, S. M.; Kim, G. B.; Kim, Y. P. Gold nanoparticle-based fluorescence quenching via metal coordination for assaying protease activity. *Gold Bull.* **2012**, *45*, 213–219.

[279] Osawa, M.; Matsuda, N.; Yoshii, K.; Uchida, I. Charge Transfer Resonance Raman Process in Surface-Enhanced Raman Scattering from p-Aminothiophenol Adsorbed on Silver: Herzberg-Teller Contribution. *J. Phys. Chem.* **1994**, *98*, 12702–12707.

[280] Ye, J.; Hutchison, J. A.; Uji-i, H.; Hofkens, J.; Lagae, L.; Maes, G.; Borghs, G.; Van Dorpe, P. Excitation wavelength dependent surface enhanced Raman scattering of 4-aminothiophenol on gold nanorings. *Nanoscale* **2012**, *4*, 1606–1611.

[281] Hong, S.; Li, X. Optimal Size of Gold Nanoparticles for Surface-Enhanced Raman Spectroscopy under Different Conditions. *J. Nanomater.* **2013**, *2013*, 1–9.

[282] Tian, F.; Bonnier, F.; Casey, A.; Shanahan, A. E.; Byrne, H. J. Surface enhanced Raman scattering with gold nanoparticles: Effect of particle shape. *Anal. Methods* **2014**, *6*, 9116–9123.

[283] Lee, C.; Robertson, C. S.; Nguyen, A. H.; Kahraman, M.; Wachsmann-Hogiu, S. Thickness of a metallic film, in addition to its roughness, plays a significant role in SERS activity. *Sci. Rep.* **2015**, *5*, 1–10.

[284] Zhao, Y.; Liu, X.; Lei, D. Y.; Chai, Y. Effects of surface roughness of Ag thin films on surface-enhanced Raman spectroscopy of graphene: Spatial nonlocality and physisorption strain. *Nanoscale* **2014**, *6*, 1311–1317.

[285] Vielchik, L. Z.; Griffith, J. P.; St. Clair, N.; Navia, M. A.; Margolin, A. L. Protein Crystals as Novel Microporous Materials. *J. Am. Chem. Soc.* **1998**, *120*, 4290–4294.

[286] Liu, X.; Wei, W.; Yuan, Q.; Zhang, X.; Li, N.; Du, Y.; Ma, G.; Yan, C.; Ma, D. Apoferritin–CeO2 nano-truffle that has excellent artificial redox enzyme activity. *Chem. Commun.* **2012**, *48*, 3155-3157.

[287] Andreescu, D.; Bulbul, G.; Özel, R. E.; Hayat, A.; Sardesai, N.; Andreescu, S. Applications and implications of nanoceria reactivity: measurement tools and environmental impact. *Environ. Sci.: Nano* **2014**, *1*, 445–458.

[288] Frey, R.; Hayashi, T.; Hilvert, D. Enzyme-mediated polymerization inside engineered protein cages. *Chem. Commun.* **2016**, *52*, 10423–10426.

[289] Heckert, E. G.; Karakoti, A. S.; Seal, S.; Self, W. T. The role of cerium redox state in the SOD mimetic activity of nanoceria. *Biomaterials* **2008**, *29*, 2705–2709.

[290] Fan, J.; Yin, J. J.; Ning, B.; Wu, X.; Hu, Y.; Ferrari, M.; Anderson, G. J.; Wei, J.; Zhao, Y.; Nie, G. Direct evidence for catalase and peroxidase activities of ferritin-platinum nanoparticles. *Biomaterials* **2011**, *32*, 1611–1618.

[291] Asati, A.; Santra, S.; Kaittanis, C.; Nath, S.; Perez, J. M. Oxidase-Like Activity of Polymer-Coated Cerium Oxide Nanoparticles. *Angew. Chem. Int. Ed.* **2009**, *48*, 2308–2312.

[292] Wang, Q.; Zhang, C.; Liu, L.; Li, Z.; Guo, F.; Li, X.; Luo, J.; Zhao, D.; Liu, Y.; Su, Z. High hydrostatic pressure encapsulation of doxorubicin in ferritin nanocages with enhanced efficiency. *J. Biotechnol.* **2017**, *254*, 34–42.

[293] Correia Carreira, S.; Armstrong, J. P. K.; Seddon, A. M.; Perriman, A. W.; Hartley-Davies, R.; Schwarzacher, W. Ultra-fast stem cell labelling using cationised magnetoferritin. *Nanoscale* **2016**, *8*, 7474–7483.

[294] Fountoulaki, S.; Daikopoulou, V.; Gkizis, P. L.; Tamiolakis, I.; Armatas, G. S.; Lykakis, I. N. Mechanistic Studies of the Reduction of Nitroarenes by NaBH4 or Hydrosilanes Catalyzed by Supported Gold Nanoparticles. *ACS Catal.* **2014**, *4*, 3504–3511.

[295] Hammett, L. P. The Effect of Structure upon the Reactions of Organic Compounds. Benzene Derivatives. *J. Am. Chem. Soc.* **1937**, *59*, 96–103.

[296] Liu, A.; Traulsen, C. H. H.; Cornelissen, J. J. L. M. Nitroarene Reduction by a Virus Protein Cage Based Nanoreactor. *ACS Catal.* **2016**, *6*, 3084–3091.

[297] Schneider, C. A.; Rasband, W. S.; Eliceiri, K. W. NIH Image to ImageJ: 25 years of image analysis. *Nat. Methods* **2012**, *9*, 671–675.

[298] Smallcombe, S. H.; Patt, S. L.; Keifer, P. A. WET Solvent Suppression and Its Applications to LC NMR and High-Resolution NMR Spectroscopy. *J. Magn. Reson.* **1995**, *303*, 295–303.

[299] Sebby, K. B.; Mansfield, E. Determination of the surface density of polyethylene glycol on gold nanoparticles by use of microscale thermogravimetric analysis. *Anal. Bioanal. Chem.* **2015**, *407*, 2913–2922.

[300] Slabu, I.; Wirch, N.; Caumanns, T.; Theissmann, R.; Krüger, M.; Schmitz-Rode, T.; Weirich, T. E. Electron tomography and nano-diffraction enabling the investigation of individual magnetic nanoparticles inside fibers of MR visible implants. *J. Phys. D. Appl. Phys.* **2017**, *50*,.

[301] Bradford, M. M. A Rapid and Sensitive Method for the Quantitation of Microgram Quantities of Protein Utilizing the Principle of Protein-Dye Binding. *Anal. Biochem.* **1976**, *72*, 248–254.

10 List of abbreviations

3AEC	3-Amino-9-ethylcarbazole
4-AP	4-Aminophenol
4-ATP	4-Aminothiophenol
4-NP	4-Nitrophenol
5-ASA	5-Aminosalicylic acid
AAS	Atomic absorption spectroscopy
AgNP	Silver nanoparticles
AgNP-Hyeon	Silver nanoparticles obtained by Hyeon *et al.* synthesis
AgNP-Yam	Silver nanoparticles obtained by Yamamoto *et al.* synthesis
a.u	Arbitrary units
AuC2	Gold nanoparticle with C2 ligand shell
AuC2mix	Gold nanoparticle with C2mix ligand shell
$AuC11^+$	Gold nanoparticle with $C11^+$ ligand shell
$AuFtn^{(neg)}$	Gold nanoparticle-loaded $Ftn^{(neg)}$
$AuFtn^{(pos)}$	Gold nanoparticle-loaded $Ftn^{(pos)}$
AuNP	Gold nanoparticle
AuNP-Leff	Gold nanoparticles obtained by Leff *et al.* synthesis
AuNP-Peng	Gold nanoparticles obtained by Peng *et al.* synthesis
C2	2-(Dimethylamino)ethanethiol
$C2^+$	(2-Mercaptoethyl)-N,N,N-trimethylammonium
C2mix	Mixed ligand shell composed of C2 and $C2^+$
$C5^+$	(5-Mercaptopentyl)-N,N,N-trimethylammonium
$C11^+$	(11-Mercaptoundecyl)-N,N,N-trimethylammonium
CD	Circular dichroism
$CeFtn^{(neg)}$	Cerium oxide nanoparticle-loaded $Ftn^{(neg)}$
$CeFtn^{(pos)}$	Cerium oxide nanoparticle-loaded $Ftn^{(pos)}$
d_{DLS}	Hydrodynamic diameter determined by DLS
d_{TEM}	Nanoparticle core size determined by TEM
DAB	3,3'-Diaminobenzidin
DCM	Dichloromethane
DDA	Dodecyl amine
DFT	Density functional theory

DIA	*o*-Dianisidine
DLS	Dynamic light scattering
DTT	Dithiothreitol
EDX	Energy-dispersive X-ray
eFtn$^{(neg)}$	Empty Ftn$^{(neg)}$
eFtn$^{(pos)}$	Empty Ftn$^{(pos)}$
FeFtn$^{(neg)}$	Iron oxide nanoparticle-loaded Ftn$^{(neg)}$
Ftn$^{(neg)}$	Negatively surface charged ferritin
Ftn$^{(pos)}$	Positively surface charged ferritin
FP	Footprint
Gua	Guanidinium hydrochloride
HEPES	2-[4-(2-Hydroxyethyl)piperazin-1-yl]ethanesulfonic acid
IEC	Ion-exchange chromatography
λ_{max}	Plasmon absorbance maxima
LSPR	Local surface plasmon resonance
MES	2-(N-Morpholino)ethanesulfonic acid
MeOH	Methanol
MWCO	Molecular weight cut-off
NB	Nitrobenzene
NBS	3-Nitrobenzenesulfonate
NMR	Nuclear magnetic resonance
NTA	N,N,N-Trimethyl-1-(3-nitrophenyl)methanaminium
OPD	*o*-Phenyldiamine
PDI	Polydispersity index
PEG	Polyethyleneglycol
RhFtn$^{(neg)}$	Rhodamine B loaded Ftn$^{(neg)}$
RhFtn$^{(pos)}$	Rhodamine 6G loaded Ftn$^{(pos)}$
SAXS	Small angle X-ray scattering
SDS	Sodium dodecyl sulfate
SEC	Size-exclusion chromatography
SEM	Scanning electron microscopy
SERS	Surface enhanced Raman spectroscopy
T_2	Spin-spin relaxation
TBAB	tert-Butylamine borane complex

TEM	Transmission electron microscopy
tetralin	1,2,3,4-Tetrahydronaphthalene
TGA	Thermogravimetric analysis
TMB	3,3',5,5'-Tetramethylbenzidine
TRIS	2-Amino-2-(hydroxymethyl)propane-1,3-diol
UV-Vis	Ultraviolet visible
XRD	X-ray diffraction
ζ-pot.	Zeta-potential

Abbreviation of amino acids

3-letter	1-letter	amino acid	3-letter	1-letter	amino acid
Ala	A	Alanine	Leu	L	Leucine
Arg	R	Arginine	Lys	K	Lysine
Asn	N	Asparagine	Met	M	Methionine
Asp	D	Aspartic acid	Phe	F	Phenylaniline
Cys	C	Cysteine	Pro	P	Proline
Gln	Q	Glutamine	Ser	S	Serine
Glu	E	Glutamic acid	Thr	T	Threonine
Gly	G	Glycine	Trp	W	Tryptophan
His	H	Histidine	Tyr	Y	Tyrosine
Ile	I	Isoleucine	Val	V	Valine

11 List of chemicals

Chemical	Supplier	H- and P statement
3-Amino-9-ethylcarbazole	Acros Organics	H: 301-350 P: 201-301+310-308+313
(3-nitrobenzyl)trimethylammonium chloride	Sigma Aldrich	H: 315-319-335 P: 261-264-271-280-302+352-304+340-305+351+338-312-321-332+313-337+313-362-403+233-405-501
5-Aminosalicylic acid	Acros Organics	H: 315-319-335 P: 261-305+351+338
4-Aminothiophenol	Sigma Aldrich	H: 314 P: 280-305+351+338-310
Ammonium iron(II) sulfate hexahydrate	Merck	-
Ammonium sulfate	AppliChem	H: 302-315-319-335-411 P: 261-264-270-271-273-280-301+312-302+352-304+340-305+351+338-312-321-330-332+313-337+313-362-391-403+233-405-501
Ampicillin sodium salt	AppliChem	H: 317-334 P: 261-272-280-284-321-302+352-304+341-333+313-342+311-362+364-501
tert-Butylamine borane complex	Sigma Aldrich	H: 301-311-312-315-319-335-411 P: 261-264-270-271-273-280-301+310-302+352-304+340-305+351+338-312-321-322-330-332+313-337+313-361-362-363-391-403+233-405-501
Cerium(III) chloride heptahydrate	Sigma Aldrich	H: 314-410 P: 305+351+338+310-280-301+330+331-303+361+353-261-302+352
Cerium oxide nanoparticles	Sigma Aldrich	H: 302 P: 264, 301+312-330-501
3,3'-Diaminobenzidin tetrahydrochloride	Alfa Aesar	H: 341-350 P: 201-308+313
o-Dianisidine dihydrochloride	Alfa Aesar	H: 302-314-318-350 P: 201-202-260-264-270-280-321-363-301+310-301+330+331-303+361+353-

		304+340-305+351+338-308+313-405-501
Dichlormethane	Fischer Scientific	H: 315-319-335-336-351-373 P: 261-281-305+351+338
2-(Dimethylamino)ethanethiol hydrochloride	Sigma Aldrich	H: 315-319-335 P: 261-264-271-280-302+352-304+340-305+351+338-312-321-332+313-337+313-362-403+233-405-501
Dithiothreitol	Sigma Aldrich	H: 302-315-319-335 P: 280-302+352-305+351+338-308+311
Dodecylamine	Sigma Aldrich	H: 304-314-335-373-410 P: 261-273-280-301+310-305+351+338-310
Ethylenediaminetetraacetic acid	Roth	H: 309 P: 305+351+338
Ethanol	Fischer Scientific	H: 225-319 P: 210-240-305+351+338-403+233
Glutaraldehyde (25%)	Merck	H: 301-330-314-317-334-335-410 P: 260-280-304+340-310-305+351+338-403+233
Glycerol	VWR	-
Gold(III) chloride trihydrate	Sigma Aldrich	H: 314-317 P: 280-305+351+338-310
Guanidinium hydrochloride	Sigma Aldrich	H: 302+332-315-319 P: 261-280-301+312-330-304+340+312-305+351+338-337+313
HEPES	Roth	H 315-319-335 P: 261-264-270-271-280-301+312-302+352-304+312-304+340-305+351+338-312-321-322-330-332+313-337+313-362-363-403+233-405-501
Hydrochloric acid (37%)	Fischer Scientific	H: 290-314-335 P: 260-280-303+361+353-304+340+310-305+351+338
Hydrogen peroxide (30%)	Chemsol	H: 271-302-314-332-335-412 P: 280-305+351+338-310
IPTG	Roth	H: 319-335-315 P: 280-302+352-304+340-305+351+338-312
Magnesium acetate tetrahydrate	Grüssing	-

Magnesium formate dihydrate	Fluka	-
(2-mercaptoethyl)-N,N,N-trimethylammonium chloride	Prochima	H: 301-302-315-317-319-335-400-410 P: 261-264-270-271-272-273-280-301+310-301+312-302+352-304+340-305+351+338-312-321-330-332+313-333+313-337+313-362-363-391-403+233-405+501
(5-Mercaptopentyl)-N,N,N-trimethylammonium chloride	Prochima	H: 301-302-315-317-319-335-400-410 P: 261-264-270-271-272-273-280-301+310-301+312-302+352-304+340-305+351+338-312-321-330-332+313-333+313-337+313-362-363-391-403+233-405+501
(11-Mercaptoundecyl)-N,N,N-trimethylammonium chloride	Prochima	H: 301-302-315-317-319-335-400-410 P: 261-264-270-271-272-273-280-301+310-301+312-302+352-304+340-305+351+338-312-321-330-332+313-333+313-337+313-362-363-391-403+233-405+501
MES monohydrate	AppliChem	H: 315-319-335 P: 261-305+351+338
Methanol	Fischer Scientific	H: 225-331-311-301-370 P: 210-233-280-302+352-304+340-308+310-403+235
Myristic acid	Sigma Aldrich	H: 315-317-318-319-411-413 P: 261-264-272-273-280-302+352-305+351+338-310-321-332+313-333+313-337+313-362-363-391-501
Nitrobenzene	Sigma Aldrich	H: 360F-301+311+331-351-372-412 P: 201-273-280-302+352-304+340-308+310
4-Nitrophenol	Sigma Aldrich	H: 301-312-332-373 P: 261-301+310-330-302+352-304+340-312
Oleic acid	Sigma Aldrich	H: 315-319-335 P: 261-264-271-280-302+352-304+340-305+351+338-312-321-332+313-337+313-362-403+233-405-501
Oleylamine (70%)	Sigma Aldrich	H: 302-304-314-335-373-410 P: 260-280-303+361+353-304+340+310-305+351+338

o-Phenyldiamine dihydrochloride	Alfa Aesar	H: 301-312-317-319-332-341-351-400-410 P: 201-202-261-264-270-271-272-273-280-281-301+310-302+352-304+312-304+340-305+351+338-308+313-312-321-322-330-333-313-337+313-363-391-405-501
di-Potassium hydrogen phosphate trihydrate	Merck	-
Potassium dihydrogen phosphate	Riedel	-
Rhodamine 6G	Acros Organics	H: 301-302-318-400-410-411 P: 264-270-273-280-301+310-301+312-305+351+338-310-321-330-391-405-501
Rhodamine B	Alfa Aesar	H: 318-412 P: 260-273-280-305+351+338
Silver nitrate	Sigma Aldrich	H: 272-290-314-410 P: 210-220-260-280-305+351+338-370+378-308-310
Sodium acetate	Grüssing	-
Sodium borohydride	Sigma Aldrich	H: 260-301-314-360F P: 201-231+232-280-308+313-370+378-402+404
Sodium chloride	Grüssing	-
Sodium citrate	Merck	-
Sodium dodecyl sulfate	Roth	H: 228-302+332-315-318-335-412 P: 210-261-280-301+312+330-305+351+338+310-370+378
Sodium hydroxide	Grüssing	H: 290-314 P: 280-301+330+331-305+351+338-308+310
Sodium-3-nitrobenzenesulfonate	Sigma Aldrich	H: 317-319 P: 280-305+351+338
Stearic acid	Sigma Aldrich	H: 315-319-335-412 P: 261-264-271-273-280-302+352-304+340-305+351+338-312-321-332+313-337+313-362-403+233-405-501
Sucrose	Fluka	-
Tetralin	Alfa Aeser	H: 304-315-319-351-411 P: 273-301+310-305+351+338-331

3,3',5,5'-Tetramethylbenzidine	Merck	H: 315-319-335 P:261-305+351+338
Toluene	Fischer Scientific	H: 225-304-315-336-361d-373 P: 210-240-301+310+330-302+352-314-403+233
TRIS	Roth	H: 315-319-335 P: 261-305+351+338
Tryptone	AppliChem	-
Uranyl acetate		H: 300+330-373-411 P: 260-264-270-271-273-284-301+310-304+340-310-314-320-321-330-391-403+233-405-501
Yeast	AppliChem	-

12 List of Figures

Figure 2.1: Nanoparticle nucleation and growth. 5

Figure 2.2: Schematic illustration of the local surface plasmon resonance. 7

Figure 2.3: Binary nanoparticle superlattices. 11

Figure 2.4: Nanoparticle assembly with DNA linker. 12

Figure 2.5: Nanoparticle assembly by DNA origami. 15

Figure 2.6: Examples of natural protein containers. 18

Figure 2.7: Ferritin protein container. 19

Figure 2.8: Interfaces in the protein container. 20

Figure 2.9: Assembly of ferritin into highly ordered structures. 23

Figure 2.10: Construction of biohybrid materials. 24

Figure 3.1: Strategy for the formation of nanoparticle superlattices with protein containers as building blocks. 27

Figure 3.2: Electrostatic surface potential of the ferritin variants. 28

Figure 3.3: Metal oxide nanoparticle synthesis in the protein container cavity. 29

Figure 3.4: Overview of different crystal structures. 29

Figure 3.5: Characterization of the nanoparticle superlattice. 30

Figure 4.1: General strategy for the construction of nanomaterials composed of plasmonic nanoparticles and protein containers used in this work. 31

Figure 4.2: Encapsulation of plasmonic nanoparticles. 32

Figure 5.1: Inner cavity of the ferritin protein container. 34

Figure 5.2: Characterization of AuNP-Leff. 36

Figure 5.3: Characterization of AuNP-Peng. 37

Figure 5.4: Ligands for ligand exchange reaction. 39

Figure 5.5: Ligand exchange reactions. 39

Figure 5.6: SERS measurement of AuC2mix. 42

Figure 5.7: ^{1}H-NMR spectra of the pure ligands. 43

Figure 5.8: Alternative strategy for the determination of the ligand shell composition by ^{1}H-NMR. 44

Figure 5.9: ^{1}H-NMR spectrum of an applied ligand ratio of 1:1. 45

Figure 5.10: Model for ligands on an AuNP surface. 47

Figure 5.11: TGA measurements of the pure ligand and AuC2mix. 47

Figure 5.12: Stability test of different functionalized AuNP in different conditions. 49

Figure 5.13: Characterization of AgNP-Yam. 52

Figure 5.14: Characterization of AgNP-Hyeon. 53

Figure 5.15: Ligand exchange reactions with AgNPs. 54

Figure 5.16: Stability tests of AgC2mix in different conditions. 55

Figure 5.17: Dis- and reassembly of the protein container. 56

Figure 5.18: Characterization of the dis- and reassembly process of $Ftn^{(pos)}$ in the Gua condition. 58

Figure 5.19: Crystallization of reassembled $Ftn^{(pos)}$. 60

Figure 5.20: Characterization of the dis- and reassembly process of $Ftn^{(neg)}$ in the Gua condition. 61

Figure 5.21: Characterization of the dis- and reassembly process of $Ftn^{(neg)}$ in the pH 2 condition. 62
Figure 5.22: Crystallization of reassembled $Ftn^{(neg)}$. 64
Figure 5.23: Crystallization of reassembled $Ftn^{(pos)}$ and $Ftn^{(neg)}$. 64
Figure 5.24: Encapsulation of AuNP inside the protein container cavity. 65
Figure 5.25: Encapsulation of different AuNPs. 66
Figure 5.26: Encapsulation of different AuNPs with the pH 2 condition. 68
Figure 5.27: Salt screening for the encapsulation of AuC2mix with the Gua condition. 70
Figure 5.28: Salt screening for the encapsulation of AuC2mix with the pH 2 condition. 71
Figure 5.29: SEC chromatogram of $AuFtn^{(pos)}$. 72
Figure 5.30: Purification of $AuFtn^{(pos)}$ by IEC. 73
Figure 5.31: Optimization of AuNP loading in $AuFtn^{(pos)}$. 74
Figure 5.32: Purification of $AuFtn^{(neg)}$ by IEC. 75
Figure 5.33: Optimization of AuNP loading in $AuFtn^{(neg)}$ with Gua condition. 76
Figure 5.34: Additional purification of $AuFtn^{(neg)}$ by SEC. 77
Figure 5.35: SEC chromatogram of $AuFtn^{(neg)}$. 78
Figure 5.36: Optimization of AuNP loading in $AuFtn^{(neg)}$ with pH 2 condition. 78
Figure 5.37: Purification of $AgFtn^{(pos)}$ by IEC. 80
Figure 5.38: Purification of $AgFtn^{(neg)}$ by IEC. 81
Figure 5.39: Encapsulation of rhodamine 6G in $Ftn^{(pos)}$. 83
Figure 5.40: Encapsulation of rhodamine B in $Ftn^{(neg)}$. 84
Figure 5.41: Synthesis of $CeFtn^{(pos)}$ and $CeFtn^{(neg)}$. 85
Figure 5.42: Synthesis of $FeFtn^{(neg)}$. 86
Figure 5.43: Assembly of the building blocks. 87
Figure 5.44: Crystallization methods. 88
Figure 5.45: Screening of Mg^{2+} concentration for unitary structure. 89
Figure 5.46: Screening of Mg^{2+} concentration for binary structure. 90
Figure 5.47: Upscaling of the batch crystallization. 91
Figure 5.48: Optical microscopy images of plasmonic nanoparticle filled protein crystals. 92
Figure 5.49: SEM measurement of $AuFtn^{(pos)}/eFtn^{(neg)}$ crystal and EDX mapping. 94
Figure 5.50: SAXS data of $AuFtn^{(pos)}/eFtn^{(neg)}$. 95
Figure 5.51: Optical microscopy image of protein crystal with different cargo. 96
Figure 5.52: Plasmin-exciton coupling in protein crystals. 98
Figure 5.53: SERS measurements of 4-ATP with different substrates. 99
Figure 5.54: Catalytic activity of $CeFtn^{(pos)}$ and $CeFtn^{(neg)}$ in solution. 102
Figure 5.55: Oxidase-like activity of CeO_2 loaded protein crystals. 104
Figure 5.56: Magnetic properties of $FeFtn^{(neg)}$. 105
Figure 5.57: Catalytic reduction of nitroarenes. 106
Figure 5.58: Catalytic reduction of 4-NP in the protein crystal. 108
Figure 8.1: ^{1}H-NMR of AuC2mix treated with aqua regia. 133
Figure 8.2: ^{1}H-NMR of DTT. 134
Figure 8.3: Disassembly of protein containers with different incubation time. 134

Figure 8.4: Disassembly of protein containers with different Gua concentrations. 135
Figure 8.5: SEC of the disassembly of $Ftn^{(pos)}$ in pH 2. 135
Figure 8.6: SEC of the reassembly of $Ftn^{(neg)}$ in pH 2. 136
Figure 8.7: TEM image of the $AuFtn^{(pos)}$ SEC. 136
Figure 8.8: UV-Vis spectra of rhodamine 6G and rhodamine B. 137
Figure 8.9: Sucrose gradient centrifugation of cerium oxide loaded protein containers. 137
Figure 8.10: Sucrose gradient centrifugation of $FeFtn^{(neg)}$. 138
Figure 8.11: Batch crystallization of $Ftn^{(neg)}$ crystals. 138
Figure 8.12: Batch crystallization of $Ftn^{(pos)}$/$Ftn^{(neg)}$ crystals. 138
Figure 8.13: SAXS data of $eFtn^{(neg)}$. 139
Figure 8.14: Crystallization of $AuFtn^{(neg)}$ obtained with the pH 2 method. 140
Figure 8.15: SERS measurement of $AuFtn^{(pos)}$/$eFtn^{(neg)}$ in water. 140
Figure 8.16: X-ray diffraction pattern of $CeFtn^{(pos)}$ 141
Figure 8.17: Oxidation of TMB. 141
Figure 8.18: Determination of kinetic parameters for oxidase-like and peroxidase-like activity. 141
Figure 8.19: UV-Vis absorption spectra of empty and CeO_2 loaded protein containers. 142
Figure 8.20: Microscopy images of oxidase-like activity of CeO_2 loaded protein crystals with TMB. . 142
Figure 8.21: Microscopy images of peroxidase-like activity of CeO_2 loaded protein crystals with TMB. 143
Figure 8.22: Catalytic cycling for oxidase-like activity. 144
Figure 8.23: Oxidase-like and peroxidase-like activity of $FeFtn^{(neg)}$. 144
Figure 8.24: Steady state kinetic for the peroxidase-like activity of $FeFtn^{(neg)}$. 145
Figure 8.25: Microscopy images of oxidase-like and peroxidase-like activity of $eFtn^{(pos)}$/$FeFtn^{(neg)}$ crystals. 146
Figure 8.26: Oxidase activity of $CeFtn^{(pos)}$/$CeFtn^{(neg)}$ crystals with OPD. 146
Figure 8.27: Oxidase activity of $CeFtn^{(pos)}$/$CeFtn^{(neg)}$ crystals with 5-ASA. 147
Figure 8.28: Oxidase activity of $CeFtn^{(pos)}$/$CeFtn^{(neg)}$ crystals with DIA. 147
Figure 8.29: Oxidase activity of $CeFtn^{(pos)}$/$CeFtn^{(neg)}$ crystals with DAB. 148
Figure 8.30: Oxidase activity of $CeFtn^{(pos)}$/$CeFtn^{(neg)}$ crystals with 3AEC. 148
Figure 8.31: UV-Vis spectra of different AuNP samples. 149
Figure 8.32: Determination of kinetic parameters for the reduction of NB. 149
Figure 8.33: Determination of kinetic parameters for the reduction of NTA. 150
Figure 8.34: Determination of kinetic parameters for the reduction of NBS. 150
Figure 8.35: Determination of the starting value. 151
Figure 8.36: Determination of the kinetic parameters. 151

13 List of Tables

Table 5.1: Summary of data for the AuNP-Leff including UV-Vis measurements and size determination by DLS and TEM. 35

Table 5.2: Summary of data for the AuNP-Peng including UV-Vis measurements and size determination by DLS and TEM. 37

Table 5.3: Summary of data for the AuNP obtained after the ligand exchange reaction including UV-Vis measurements, size determination by DLS and ζ-pot. measurements. 41

Table 5.4: ζ-pot. of AuC2mix with different used compositions for the ligand exchange. 41

Table 5.5: Summary of the effective ligand ratio on the AuNP as determined by NMR for four different applied ligand ratios for the ligand exchange reaction. 45

Table 5.6: Size of the protein container before and after reassembly measured by DLS and TEM. 59

Table 5.7: Size of the protein container before and after reassembly determined by DLS and TEM. 63

Table 5.8: DLS, TEM and ζ-pot. data for AuFtn$^{(pos)}$, Ftn$^{(pos)}$ and AuNP. 74

Table 5.9: DLS, TEM and ζ-pot. data for AuFtn$^{(neg)}$ and Ftn$^{(neg)}$. 76

Table 5.10: Cell parameters obtained from SAXS data. 95

Table 5.11: Comparison of the position of Raman bands in different measurements. 100

Table 5.12: Comparison of kinetic parameters for the oxidase-like and peroxidase-like activity. 102

Table 5.13: Reaction rates of the reduction of substituted benzene derivates with different AuNP samples. 107

Table 7.1: List of used strains. 113

Table 7.2: Buffers used for the purification of the protein variants, the synthesis of nanoparticles and dis- and reassembly of the protein containers. 114

Table 7.3: Different gold to ligand ratio for the optimization of the AuNP size. 124

Table 8.1: List of mutations in Ftn$^{(pos)}$ and Ftn$^{(neg)}$. 133

Table 8.2: Cell parameters obtained from SAXS data. 139

Table 8.3: Kinetic parameters of the peroxidase-like activity of FeFtn$^{(neg)}$. 145

Acknowledgment - Danksagung

Ich danke meinem Doktorvater Prof. Tobias Beck für die Betreuung dieser Arbeit und der Möglichkeit an diesem spannenden Projekt zu arbeiten. Ich möchte mich insbesondere bedanken für seine umfangreiche Hilfe, die zahlreichen Diskussionen, die wissenschaftlichen Freiheiten sowie für die Gelegenheiten, an nationalen und internationalen Tagungen teilnehmen zu können.

Prof. Alf Mews danke ich für die freundliche Übernahme des Gutachtens für die Dissertation und Prof. Christian Betzel, Prof. Axel Jacobi von Wangelin und Prof. Holger Lange als Mitglied der Prüfungskommission.

Dem Cusanuswerk danke ich für die finanzielle und ideelle Förderung.

Prof. Ulrich Simon danke ich für den Zugang zu seinen Laboren und Geräten, der Finanzierung, sowie der Möglichkeit an der Teilnahme an den Gruppenseminaren.

Des Weiteren danke ich Prof. Ulrich Schwanenberg für den Zugang zu seinen Laboren und Geräten.

Bei meinen Arbeitskollegen Dr. Matthias Künzle, Made Budiarta, Michael Rütten, Hendrik Böhler und Dr. Brandon Seychell bedanke ich mich für die freundschaftliche Arbeitsatmosphäre und für die zahlreichen wissenschaftlichen Diskussionen. Insbesondere möchte ich mich auch bei allen anderen Doktoranden im Labor 002, besonders Dr. Martin Davi, Ricarda Pütt und Dr. Oliver Linnenberg, für die schöne Zusammenarbeit bedanken. Dem AK Simon danke ich für die nette Arbeitsatmosphäre und freundliche Unterstützung.

Außerdem möchte ich meinen Bachelorstudenten Robert Rauschen und meinen Forschungsstudenten Esther Jaeckel, Barbara Bong, Marcus Lehnertz und Markus Ottersbach, die mit ihren Arbeiten direkt oder indirekt zum Gelingen dieser Dissertation beigetragen haben, danken.

Beim AK Mews bedanke ich mich für die zahlreiche und freundliche Hilfe bei allen möglichen Problemen und Fragen.

Mein Dank geht an Dr. Andreas Meyer vom Institut für physikalische Chemie in Hamburg für die SAXS Messungen der Proteinkristalle und Dr. Ioana Slabu vom Lehrstuhl für angewandte Medizintechnik in Aachen für die magnetische Charakterisierung der eisenoxidbeladenen Proteincontainern.

Aus dem Institut für anorganische Chemie in Aachen danke ich für die Durchführung von Messungen: Sophia Peter (Raman), Dr. Gerhard Fink (NMR), Björn Faßbender (AAS) und Brigitte Jansen (TGA). Des Weiteren möchte ich mich für durchgeführte Messungen am Institut für physikalische Chemie in Hamburg bei Robert Schön (SEM) und Roman Kusterer (Emissionsspektren der Proteinkristalle) bedanken.

Meinen Freunden aus der Zeit in Aachen Nico, Lars, Lucas, Artur, Max, Julian und Wolle, sowie aus der Heimat Max, Anna, Dennis, Elena, Sabrina, Eric, Julia und Maxi, danke ich für eine tolle gemeinsame Zeit außerhalb der Promotion.

Zum Schluss geht mein besonderer Dank meinen Eltern, meinem Bruder Michael und meiner Schwester Melanie, auf deren Unterstützung, Verständnis und Hilfe ich jeder Zeit zählen kann.

Curriculum Vitae

Personal

Name:	Marcel Josef Lach
Date of Birth:	17.11.1991
Place of Birth:	Düsseldorf
Nationality:	German

Education

01/2020 – 09/2020	**Doctoral studies** Institute of Physical Chemistry Universität Hamburg
10/2016 – 12/2019	**Doctoral studies** Institute of Inorganic Chemistry RWTH Aachen
10/2014 – 09/2016	**Master of Science in Chemistry** RWTH Aachen
10/2011 - 09/2014	**Bachelor of Science in Chemistry** RWTH Aachen
08/2002 – 07/2011	**Allgemeine Hochschulreife** Geschwister-Scholl Gymnasium

Ich habe fertig.

\- Giovanni Trapattoni -

www.ingramcontent.com/pod-product-compliance
Ingram Content Group UK Ltd.
Pitfield, Milton Keynes, MK11 3LW, UK
UKHW021652190726
13853UKWH00001B/215

9 783736 972964